Multi-Pulse Evolution and Space-Time Chaos in Dissipative Systems

MEMOIRS
of the
American Mathematical Society

Number 925

Multi-Pulse Evolution and Space-Time Chaos in Dissipative Systems

Sergey Zelik
Alexander Mielke

March 2009 • Volume 198 • Number 925 (second of 6 numbers) • ISSN 0065-9266

American Mathematical Society
Providence, Rhode Island

2000 *Mathematics Subject Classification.*
Primary 35Q30, 37L30.

Library of Congress Cataloging-in-Publication Data

Zelik, Sergey, 1972–
Multi-pulse evolution and space-time chaos in dissipative systems / Sergey Zelik, Alexander Mielke.
p. cm. — (Memoirs of the American Mathematical Society, ISSN 0065-9266 ; no. 925)
"Volume 198, number 925 (second of 6 numbers)."
Includes bibliographical references.
ISBN 978-0-8218-4264-5 (alk. paper)
1. Attractors (Mathematics) 2. Lyapunov exponents. 3. Stokes equations. I. Mielke, Alexander, 1958– II. Title.

QA614.813.Z45 2009
515′.39—dc22 2008047916

Memoirs of the American Mathematical Society

This journal is devoted entirely to research in pure and applied mathematics.

Subscription information. The 2009 subscription begins with volume 197 and consists of six mailings, each containing one or more numbers. Subscription prices for 2009 are US$709 list, US$567 institutional member. A late charge of 10% of the subscription price will be imposed on orders received from nonmembers after January 1 of the subscription year. Subscribers outside the United States and India must pay a postage surcharge of US$65; subscribers in India must pay a postage surcharge of US$95. Expedited delivery to destinations in North America US$57; elsewhere US$160. Each number may be ordered separately; *please specify number* when ordering an individual number. For prices and titles of recently released numbers, see the New Publications sections of the *Notices of the American Mathematical Society.*

Back number information. For back issues see the *AMS Catalog of Publications.*

Subscriptions and orders should be addressed to the American Mathematical Society, P. O. Box 845904, Boston, MA 02284-5904, USA. *All orders must be accompanied by payment.* Other correspondence should be addressed to 201 Charles Street, Providence, RI 02904-2294, USA.

Memoirs of the American Mathematical Society (ISSN 0065-9266) is published bimonthly (each volume consisting usually of more than one number) by the American Mathematical Society at 201 Charles Street, Providence, RI 02904-2294, USA. Periodicals postage paid at Providence, RI. Postmaster: Send address changes to Memoirs, American Mathematical Society, 201 Charles Street, Providence, RI 02904-2294, USA.

This publication is indexed in *Science Citation Index*®, *SciSearch*®, *Research Alert*®, *CompuMath Citation Index*®, *Current Contents*®*/Physical, Chemical & Earth Sciences.*
Printed in the United States of America.

The paper used in this book is acid-free and falls within the guidelines
established to ensure permanence and durability.
Visit the AMS home page at http://www.ams.org/

10 9 8 7 6 5 4 3 2 1 14 13 12 11 10 09

Contents

Abstract

We study semilinear parabolic systems on the full space $\mathbb{R}^n$ that admit a family of exponentially decaying pulse-like steady states obtained via translations. The multi-pulse solutions under consideration look like the sum of infinitely many such pulses which are well separated. We prove a global center-manifold reduction theorem for the temporal evolution of such multi-pulse solutions and show that the dynamics of these solutions can be described by an infinite system of ODEs for the positions of the pulses.

As an application of the developed theory, we verify the existence of Sinai-Bunimovich space-time chaos in 1D space-time periodically forced Swift-Hohenberg equation.

Received by the editor February 2, 2006.

2000 *Mathematics Subject Classification.* 35Q30, 37L30.

Key words and phrases. dissipative systems, unbounded domains, multi-pulses, normal hyperbolicity, center-manifold reduction, space-time chaos, Bernoulli shifts.

This work is partially supported by Alexander von Humboldt foundation and by the CRDF grant RUM1-2654-MO-05. The authors are also grateful to A.Vladimirov and D.Turaev for stimulating discussions.

1. Introduction

It is well-known that even relatively simple dynamical systems generated by low-dimensional systems of ODEs can demonstrate very complicated irregular behavior (the so-called deterministic chaos). It is also well-known that the temporally localized solutions (pulses, homoclinic loops) play a crucial role in the modern theory of the deterministic chaos, see [**KaH95**] and references therein. The main feature that makes these structures so important is the ability to construct more complicated multi-pulse structures by taking the appropriate sums of "shifted" pulses. In particular, very often the existence of a *single* pulse allows to find the huge family of multi-pulses parameterized by the elements of the Bernoulli scheme $\mathcal{M}^1 := \{0,1\}^{\mathbb{Z}}$ such that the evolution operator will be conjugated with the shift operator on $\mathcal{M}^1$. Thus, the existence of a *single* (transversal) pulse allows to verify e.g., that the topological entropy of the system is strictly positive and to obtain the description of the deterministic chaos in terms of Bernoulli shifts on $\mathcal{M}^1$.

The dynamical behavior of infinite-dimensional systems generated by evolution partial differential equations are usually much more complicated (and essentially less understood), since, in addition to irregular temporal dynamics, the formation of complicated *spatial* patterns usually takes place (the so-called spatial chaos phenomena) and, as a result of irregular evolution of spatially-chaotic structures, the so-called *space-time* chaos may appear. These phenomena are genuinely infinite-dimensional and can demonstrate the essentially higher level of complexity which is not observable in the classical finite-dimensional theory, see e.g. [**CoE99a, Zel03b, Zel04**] for the case of dissipative dynamics in unbounded domains.

Nevertheless, analogously to the finite-dimensional case, localized structures are very important for the understanding the space-time dynamics generated by PDEs. Moreover, in contrast to the finite-dimensional case, here we have the additional spatial variables, such that *space*, *time* or *space-time* localized solutions can be a priori considered. The most studied and, in a sense, most interesting is the case of space-localized structures, in particular, spatially localized equilibria or traveling solitons or multi-solitons, especially in the case of integrable Hamiltonian systems, where these objects can be completely described by the inverse scattering methods.

However, the existence of soliton-like solutions is not restricted by integrable systems and takes place for many PDEs which are far from being integrable, e.g., for the so-called *dissipative* systems including the classical pattern formation equations like Ginzburg-Landau, Swift-Hohenberg, Cahn-Hilliard equations etc. Thus, spatially localized structures are one of the typical phenomena that arise in spatially extended systems under the pattern formation including hydro- and magneto-hydrodynamics, quantum physics, mathematical biology etc., see [**AfM01, BGL97, BlW02, CoM03, Cou02, DGK98, PeT01, REW00, San02**].

On the other hand, analogously to the homoclinic loops in ODEs, spatially localized solutions (=spatial pulses) can be considered as elementary building blocks for constructing more complicated multi-pulse structures by taking a sum of spatially well-separated pulse solutions and describing the space-chaos phenomena, see [**ABC96, AfM01, Ang87, Bab00, DFKM96, MiH88, PeT01, Rab93, REW00, San93**]. Furthermore, even if the initial spatial pulse is independent of time, for general pulse configurations the associated multi-pulse solution is not an equilibrium due to the weak "tail" interaction between pulses, but evolves slowly in time (in particular, this evolution can be temporarily chaotic and even produce the so-called Sinai-Bunimovich space-time chaos as it is shown below on a model example of 1D Swift-Hohenberg equation with space-time periodic external forcing). Thus, the ability to give an effective description of these weak tail interactions between pulses becomes crucial for the understanding of multi-pulse structures and their evolution.

On the physical level of rigor, the required slow evolution equations for multi-pulse structures can be relatively easily obtained by inserting the multi-pulse configuration $u(t,x)$ in the form

$$u(t,x) = \sum_i V(x - \xi_i(t)) + \text{ ``small''}$$

(where $V = V(x)$ is the initial pulse and $\xi_i(t)$ is the position of the center of the i-th pulse at time t) into the equation considered and making the asymptotic expansion with respect to the large parameter $L = \inf_{i\neq j} \|\xi_i - \xi_j\|$. Then, dropping out the (formally) higher terms in this expansion, one arrives at a reduced system of ODEs for the pulse centers $\xi_i(t)$.

We however note that, although the above expansions are widespread in the physical literature (see e.g. [**Cou02, SkV02, TVM03, VFK*99, VKR01, VMSF02**]), the rigorous justification of this reduction (based on the center manifold or the Liapunov-Schmidt reduction technique) is a highly nontrivial mathematical problem which is solved only for rather particular cases.

Indeed, to the best of our knowledge up to the moment the general center manifold reduction was obtained only for the finite number of pulses (or for space-periodic pulse configurations) and only in the case of one spatial direction, see [**Ei2002, San93, San02**]. Moreover, even in that case, only the *local* center manifold reduction theorem (in a small neighborhood of a *fixed* pulse configuration) is available in the literature, see [**Ei2002**]. This theorem is mainly adapted for the study of the multi-pulse equilibria and their stability and is clearly insufficient for describing the multi-pulse evolution if the pulses undergo large changes of their position. A global multi-pulse center manifold theorem with one manifold for all admissible pulse-configurations (also for finite number of pulses and 1D case) is announced to be proven in [**Sand**]. We also note that for the particular case of multi-kinks in 1D Chafee-Infante equations the infinite number of kinks has been considered in [**EcR98**].

The main aim of the present paper is to give a systematic study of the above justification problem in the case of parabolic dissipative systems in $\mathbb{R}^n$ for an infinite number of pulses. Moreover, keeping in mind the applications to the space-time chaos problem, we study also space-time periodic perturbations of the multi-pulse structures.

To be more precise, we consider the following evolutionary problem in $x \in \mathbb{R}^n$:

$$\partial_t u + A_0 u + \Phi(u) = \mu R(t,x,u) \tag{1.1}$$

where $u = (u^1, \cdots, u^m)$ is the unknown vector-valued function, A_0 is a positively defined uniformly elliptic operator of order $2l$ in $[L^2(\mathbb{R}^n)]^m$ with constant coefficients. $\Phi(u) := \Phi(u, D_x u, \cdots, D_x^{2l-1}u)$ is a nonlinear interaction function which is assumed to be smooth, with $\Phi(0) = \Phi'(0) = 0$. $R(t,x,u) = R(t,x,u,\cdots,D_x^{2l-1}u)$ is a perturbation which is also assumed to be smooth with respect to $u, \cdots, D_x^{2l-1}u$, and μ is a small perturbation parameter. We consider this equation in the phase space $\mathbb{X}_b$ of spatially bounded functions (and do not assume any decay conditions as $|x| \to \infty$). To be more precise, $\mathbb{X}_b = W_b^{2l(1-1/p),p}(\mathbb{R}^n)$ with large p (or $\mathbb{X}_b = C_b^{2l-1}(\mathbb{R}^n)$), see Section 3 for rigorous definitions.

Our basic assumption is that the unperturbed equation

$$\partial_t u + A_0 u + \Phi(u) = 0 \tag{1.2}$$

possesses a pulse equilibrium $V = V(x)$, $|V(x)| \le C\,\mathrm{e}^{-\alpha|x|}$, for some $\alpha > 0$. Thus, we assume that the initial spatially localized structure is a priori given and will concentrate ourselves on the study of the multi-pulse structures generated by V.

We recall that equation (1.2) is spatially homogeneous and consequently possesses a group of spatial shifts $\{T_\xi, \xi \in \mathbb{R}^n\}$, $(T_\xi u)(x) := u(x-\xi)$ as a symmetry group. Moreover, in addition to these shifts equation (1.2) is often invariant with respect to some rotations $\gamma := (\gamma_1, \gamma_2) \in SO(n) \times SO(m)$ that generate a linear action on $[C^\infty(\mathbb{R}^n)]^m$ via

$$(\mathcal{T}_\gamma u)(x) := \gamma_2 u(\gamma_1 x). \tag{1.3}$$

In order to take into account this typical situation in our general theory, we assume that equation (1.2) is invariant with respect to the action of a Lie group G of symmetries which is a skew product of the shifts $\mathbb{R}^n$ and some (compact) subgroup $\bar{G}$. For simplicity, this compact Lie group is assumed to be a subgroup of $SO(n) \times SO(m)$ (see Section 2 for slightly more general symmetry groups). Thus, the Lie group of symmetries $\{\mathcal{T}_\Gamma, \Gamma \in G\}$ of equation (1.2) is also assumed to be a priori given.

Therefore, the initial pulse is not isolated, but generates the whole manifold $\mathbb{P}_1$ of shifted and rotated pulses

$$\mathbb{P}_1 := \{V_\Gamma := \mathcal{T}_\Gamma V, \quad \Gamma \in G\} \tag{1.4}$$

(the so-called one-pulse manifold) parameterized by the elements Γ of the symmetry group G. To be more precise, $\mathbb{P}_1$ is a finite-dimensional submanifold of the phase space diffeomorphic to the factor manifold $G/\operatorname{St}_G(V)$, where $\operatorname{St}_G(V)$ is a stabilizer of V in G. In order to avoid the technicalities, we assume that this stabilizer is trivial

$$\operatorname{St}_G(V) = \{\mathrm{Id}\}. \tag{1.5}$$

and, consequently, $\mathbb{P}_1$ is diffeomorphic to G (we emphasize that G is not necessarily the *whole* group of symmetries of (1.2), so (1.5) does not seem to be a great restriction and is satisfied in many interesting applications).

Finally, we assume that the invariant one-pulse manifold $\mathbb{P}_1$ is normally-hyperbolic with respect to equation (1.2), i.e., in the exponential trichotomy $(\mathcal{V}_+, \mathcal{V}_0, \mathcal{V}_-)$ for linearized equation

$$\partial_t v + A_0 v + \Phi'(V) v = 0 \tag{1.6}$$

(which always exists due to the Fredholm property, see Section 2) the neutral subspace $\mathcal{V}_0$ coincides with the tangent plane $T_V\mathbb{P}_1$ to the manifold $\mathbb{P}_1$ at V:

$$\mathcal{V}_0 = T_V\mathbb{P}_1. \tag{1.7}$$

We recall that the neutral subspace is always finite dimensional (Fredholm property) and the tangent space $T_V\mathbb{P}_1$ is always contained in it, so assumption (1.7) can be considered as a non-degeneracy assumption (= minimal degeneracy assumption) and is also natural if the initial pulse is not "bifurcating".

It worth to mention here that, although the normal hyperbolicity assumption is postulated at one point V of the one-pulse manifold $\mathbb{P}_1$, due to the symmetry group action on $\mathbb{P}_1$, it holds *automatically* at every point $V_\Gamma \in \mathbb{P}_1$ and the trichotomy constants will be also *uniform* with respect to all points of $\mathbb{P}_1$. In a fact, only this uniform normal hyperbolicity is crucial for our method and the symmetry group structure of the manifold $\mathbb{P}_1$ is postulated in (1.4) *only* in order to avoid the technicalities and simplify the verification of that uniform normal hyperbolicity, see Remark 2.5 below.

We are now able to introduce the main object of the paper – the multi-pulse manifold

$$\mathbb{P}(L) := \{\mathbf{m} = \sum_{j=1}^{\infty} V_{\Gamma_j},\ \ \Gamma_j := (\xi_j, \gamma_j) \in G,\ \ \inf_{i\neq j} \|\xi_i - \xi_j\|_{\mathbb{R}^n} > 2L\} \tag{1.8}$$

which describes all possible multi-pulse configurations consisting of well-separated ($2L$-separated) pulses. We prove in Section 4 that, for sufficiently large L, $\mathbb{P}(L)$ is, indeed, a submanifold of the phase space with the boundary

$$\partial\mathbb{P}(L) := \{\mathbf{m} \in \mathbb{P}(L-\varepsilon),\ \ \varepsilon > 0 \quad \sup_{i,j\in\mathbb{N},\, i\neq j} \|\xi_i - \xi_j\| = 2L\}.$$

Thus, a multi-pulse configuration is determined by well-separated sequence $\{\Gamma_j\}_{j=1}^{\infty}$ of symmetry group elements and, therefore, its time evolution can be described by symmetry group-valued functions $\{\Gamma_j(t)\}_{j=1}^{\infty} \in G$.

The following theorem, which gives the center manifold reduction for equation (1.1) near the manifold $\mathbb{P}(L)$ to the appropriate equation on $\mathbb{P}(L)$ can be considered as the first main result of the paper.

THEOREM 1.1. *Let the above assumptions hold and let $l \in \mathbb{N}$ be arbitrary. Then, for a sufficiently large L and sufficiently small perturbation parameter μ, there exist a time-dependent vector field $\mathrm{f}(t,\cdot)$ on $\mathbb{P}(L)$ and a function $\mathbb{W}(t,\cdot) : \mathbb{P}(L) \to \mathbb{X}_b$ which are C^l-smooth and uniformly small, viz.,*

$$\|\mathrm{f}(t,\cdot)\|_{C^l} + \|\mathbb{W}(t,\cdot)\|_{C^l} \le C(\mathrm{e}^{-\alpha L} + \mu) \tag{1.9}$$

and satisfy the following properties:

1) Every solution $\mathbf{m}(t)$ of the reduced equation

$$\frac{d}{dt}\mathbf{m}(t) = \mathrm{f}(t, \mathbf{m}(t)) \tag{1.10}$$

belonging to $\mathbb{P}(L+1) \subset \mathbb{P}(L)$ for any $t \in [\tau, T]$, generates a multi-pulse solution of the initial problem (1.1) via

$$u(t) := \mathbf{m}(t) + \mathbb{W}(t, \mathbf{m}(t)),\quad t \in [\tau, T]. \tag{1.11}$$

2) Vice versa, every solution u of (1.1) which is close to the manifold $\mathbb{P}(L)$ for all $t \in \mathbb{R}$ can be represented in the form (1.11) where $\mathbf{m}(t)$ is an appropriate solution of the reduced equation (1.10).

The reduced equations (1.10) can be rewritten in "coordinates" $\Gamma_j(t)$ as follows

$$\frac{d}{dt}\Gamma_j(t) = \tilde{\mathrm{f}}_j(t, \vec{\Gamma}),\quad j \in \mathbb{N}, \tag{1.12}$$

where $\vec{\Gamma} := \{\Gamma_j\}_{j=1}^{\infty}$ and, consequently, gives the infinite system of ODEs describing the weak temporal evolution in the well-separated multi-pulse structures.

Moreover, we give a simple formula for the leading terms in the asymptotic expansion of the vector field f as $L \to \infty$ and compute them explicitly for a number of equations of mathematical physics including the 1D Ginzburg-Landau and Swift-Hohenberg equations, see Section 10. In particular, let us consider the case of Swift-Hohenberg equation

$$\partial_t u + (\partial_x^2 + 1)^2 u + \beta^2 u + f(u) = 0, \quad \beta \neq 0. \tag{1.13}$$

Since the equation is one-dimensional and scalar, we do not have here any rotations and the symmetry group coincides with the 1D shifts $G = \mathbb{R}$. We assume the existence of the symmetric normally hyperbolic pulse equilibrium V, $V(x) = V(-x)$, for that equation (see [**BGL97**] and [**GlL94**] for the sufficient conditions which give its existence). Thus, the multi-pulse configuration is determined now by the sequence $\{\xi_j\}_{j=-\infty}^{\infty} \in \mathbb{R}$, $\xi_{j+1} - \xi_j > 2L$ of the positions of pulse centers. Then, the reduced pulse-interaction equations read:

$$\begin{aligned}\frac{d}{dt}\xi_j = M_0[&\mathrm{e}^{-\alpha(\xi_j-\xi_{j-1})}\sin(\omega(\xi_j-\xi_{j-1})+\phi_0)- \\ &-\mathrm{e}^{-\alpha(\xi_{j+1}-\xi_j)}\sin(\omega(\xi_{j+1}-\xi_j)+\phi_0)] + O(\mathrm{e}^{-4(\alpha-\varepsilon)L}),\end{aligned} \tag{1.14}$$

where $\lambda = \alpha + i\omega$ solves $(\lambda^2+1)^2 + \beta^2 = 0$ and the amplitude $M_0 \neq 0$ and phase ϕ_0 are some constants depending on the nonlinearity f, but are independent of L.

We are also interested in the dynamics of (1.1) near the constructed center manifold. In particular, in the spectrally stable case (i.e., where the subspace $\mathcal{V}_+$ in the trichotomy for (1.6) is trivial) we verify the exponential convergence to the manifold and the existence of an asymptotic phase.

THEOREM 1.2. *Let the assumptions of previous theorem hold and let, in addition, the pulse V be spectrally stable. Then, for every initial data $u(\tau) = u_\tau$ close to the manifold $\mathbb{P}(L)$, the corresponding solution $u(t)$ of* (1.1) *exists and remains close to that manifold for $t \geq \tau$ as long as it remains far from the boundary $\partial\mathbb{P}(L)$. Moreover, there exists a solution $u_0(t)$ in the form* (1.11) *for the appropriate solution* $\mathbf{m}$ *of* (1.10) *such that*

$$\|u(t) - u_0(t)\|_{\mathbb{X}_b} \leq C\,\mathrm{e}^{-\gamma(t-\tau)}, \quad \gamma > 0$$

where the constants C and γ are independent of the concrete choice of the initial data u_τ, t and τ. Thus, the dynamics of (1.1) *near $\mathbb{P}(L)$ is completely determined by reduced equations* (1.10).

Furthermore, keeping in mind the applications to the space-time chaos problem, we have also verified the continuity of the functions f and $\mathbb{W}$ in the local topology $C^{2l-1}_{\mathrm{loc}}(\mathbb{R}^n)$ and studied the relations between hyperbolic trajectories of the whole and of the reduced systems and their stability, see Section 9.

Although we consider in the paper only the case of whole physical space $\Omega = \mathbb{R}^n$, under the minor changes the method is applicable for any sufficiently large regular domain Ω if we assume that the pulses are situated sufficiently far from the boundary (see also [**BaS02**] and references therein for the case of pulses localized near the boundary). Furthermore, one can also consider the structures generated by *different* pulses and even with different symmetry groups. Moreover, one can relax also assumption (1.7) and consider multi-pulse structures with bifurcating pulses. In this case, together with the global symmetry group variables Γ the "state" of the pulse will be described also with the additional "small" bifurcation variables associated with $\mathcal{V}_0/T_V\mathbb{P}_1$.

Finally, the developed methods are also applicable for the case where the initial spatially localized structure is time-dependent, e.g. traveling wave solutions or, more general, spatially localized solutions with hyperbolic (chaotic) temporal dynamics. We return to that in the subsequent work.

The rest of the paper is devoted to the application of the above developed theory to the problem of detecting and describing space-time chaos in the concrete equations of mathematical physics. We recall that, in spite of a huge amount of numerical and experimental data on various types of space-time irregular and turbulent behavior in various physical systems, see e.g. [**GEP98, Man90, Man95, REW00**] and references therein, there are very few rigorous mathematical results on this topic and mathematically relevant models describing this phenomenon.

The most simple and natural known model for this phenomenon is the so-called Sinai-Bunimovich space-time chaos which was initially defined and found for discrete lattice dynamics, see [**AfF00, BuS88, PeS88, PeS91**]. We also recall that this model consists of a $\mathbb{Z}^n$-grid of temporally chaotic oscillators coupled by a weak interaction. Then, if the single chaotic oscillator of this grid is described by the Bernoulli scheme $\mathcal{M}^1 := \{0,1\}^{\mathbb{Z}}$, the uncoupled system naturally has an infinite-dimensional hyperbolic set homeomorphic to the multi-dimensional Bernoulli scheme $\mathcal{M}^{n+1} := \{0,1\}^{\mathbb{Z}^{n+1}} = (\mathcal{M}^1)^{\mathbb{Z}^n}$. The temporal evolution operator is then conjugated to the shift in $\mathcal{M}^{n+1}$ along the first coordinate vectors and the other n coordinate shifts are associated with the spatial shifts on the grid. Finally, due to the stability of hyperbolic sets, the above structure survives under sufficiently small coupling. Thus, according to this model, the space-time chaos can be naturally described by the multi-dimensional Bernoulli scheme $\mathcal{M}^{n+1}$.

It is worth to note that although this model is clearly not relevant for describing the space-time chaos in the so-called fully developed turbulence (since it does not reproduce the typical properties like energy cascades and Kolmogorov's laws which are believed to be crucial for understanding of this phenomenon), it can be useful and relevant for describing weak space-time chaos and weak turbulence (close to the threshold) where the generation and long-time survival of such global spatial patterns is still possible.

Unfortunately, even for this simplest model, the verification of such space-time chaos in continuous media described by PDEs occurs to be an extremely complicated problem. Moreover, even the existence of a single PDE that possesses an infinite-dimensional Bernoulli scheme was a long-standing open problem. The first examples of such PDEs (in the class of reaction-diffusion systems), have been recently constructed in [**MiZ04**]. However, the method used in that paper is based on a direct modulation of a spatial grid by the structure of the nonlinearity and leads to a very special (and rather artificial) nonlinear interaction function, which is far from the classical nonlinearities arising in physical models.

In this paper we present an alternative method of finding Sinai-Bunimovich space-time chaos in continuous media based on the multi-pulse center manifold reduction theorem formulated before. Indeed, reduced equations (1.12) for the multi-pulse evolution have the form of spatially discretized dynamical system close to dynamical systems on lattices. So, the most singular and complicated passing from spatially continuous to spatially discrete dynamics is already done and the problem is reduced to the following one: Find a spatial pulse configuration in such way that the reduced equations (1.12) would have the form of weakly coupled system of chaotic oscillators situated at the nodes of a some grid. Applying then the Sinai-Bunimovich theory to the reduced equations, we obtain the existence of the above space-time chaos in the initial equation (1.1).

However, the initial *standing* pulse is not a very convenient object for the realization of this scheme, since it does not have its own dynamics and all dynamics appears due to the coupling, so the coupling cannot be "small" with respect to this dynamics as in Sinai-Bunimovich model. In order to overcome this difficulty, we suggest to find a pulse-pattern consisting of *finite* number of pulses which demonstrate the temporally chaotic (hyperbolic) dynamics and treat it as a single temporally chaotic pulse. If we consider then a grid of well-separated temporally chaotic pulses, its reduced dynamics would be governed by a lattice model of Sinai-Bunimovich type. Of course, if the temporally chaotic pulse is a priori known, the situation becomes much simpler (see e.g., [**ASCT01, BlW02, TVZ07**] for examples of such pulses). Nevertheless, the direct verification and investigation of the existence of chaotic pulses is much more delicate problem than the analogous one for the standing pulse. So, the above construction of a chaotic pulse from standing ones seems to be applicable to a wider class of equations. In particular, the existence of pulse solutions is known for a large class of so-called modulation equations (like Swift-Hohenberg, Ginzburg-Landau equation, etc., see e.g. [**Mie02**]). So, if hyperbolic space-chaos will be detected in such equations, it can be "lifted" to many other physically relevant equations (including even Navier-Stokes and other hydrodynamical equations) using the spatial center manifold reduction, see [**Kir82, Kir85, Mie86, Mie02**]. In particular, we refer to [**AfM01**] for the existence of multi-pulse solutions in the Poiseuille problem which is based on the corresponding result in CGL equation of [**AfM99**].

In order to avoid the technicalities, we realize the above scheme only for the case of space-time periodically 1D Swift-Hohenberg equation

$$\partial_t u + (\partial_x^2 + 1)^2 u + \beta^2 u + f(u) = \mu h(t,x), \quad f(u) = u^3 + \kappa u^2, \quad \mu \ll 1 \tag{1.15}$$

(the space-time chaos of Ginzburg-Landau equations and infinite dimensional hyperbolic local *attractors* for the Swift-Hohenberg equations will be considered in the forthcoming papers [**TZ07b**] and [**TZ07a**]).

The following theorem, which establishes the existence of Sinai-Bunimovich space-time chaos for the Swift-Hohenberg equations, can be considered as the second main result of the paper.

THEOREM 1.3. *There exists a nonempty open set of parameters $(\beta, \kappa) \in \mathbb{R}^2$ of equation* (1.15) *such that, for every (β, κ) belonging to this set and every sufficiently small $\mu \neq 0$ there exists a space-time periodic external force h, $\|h\|_{C_b(\mathbb{R}^2)} \le 1$ such that the associated equation* (1.15) *possesses an infinite-dimensional hyperbolic set homeomorphic to the Bernoulli shifts $\mathcal{M}^2$ such that the Bernoulli shifts on $\mathcal{M}^2$ are conjugated to the space-time shifts (scaled by the space and time periods of h) on the above hyperbolic set.*

The reason of adding the space-time periodic perturbation into the equation considered is two-fold:

1) Swift-Hohenberg related: It is known that equation (1.15) with $\mu = 0$ is so-called extended gradient system which cannot have the space-time chaos, see [**Zel04**], so at least a temporally non-autonomous perturbation is unavoidable here.

2) Absence of hyperbolicity (structural stability): The invariance with respect to the continuous groups of space and time shifts in space or space-time homogeneous systems in unbounded domains (see Remark 9.2 below) leads to new essential difficulties.

Nevertheless, in the forthcoming paper [**TZ07b**], we will partially extend Theorem 1.3 to the case of space-time homogeneous Ginzburg-Landau equation (of course by relaxing the hyperbolicity assumption).

The paper is organized as follows. The rigorous assumptions on the structure of the considered equation, its a priori given spatial pulse V and the symmetry group G, and the immediate corollaries of these assumptions are discussed in Section 2. In Section 3 we formulate and prove a number of results on the regularity of solutions of parabolic problems in weighted Sobolev spaces that are the main technical tools of the paper.

The geometric structure of the multi-pulse manifold $\mathbb{P}(L)$ and its relations with the original problem (1.1) is studied in Sections 4, 5 and 6. In particular, we prove here that $\mathbb{P}(L)$ is indeed a submanifold of the phase space consisting of "almost-equilibria" of equation (1.1) (Section 4). We construct a uniform family of projectors $\mathbb{P}_{\mathbf{m}}$ to the tangent space $T_{\mathbf{m}}\mathbb{P}(L)$, $\mathbf{m} \in \mathbb{P}(L)$ (Section 5) and study the general structure of differential equations on $\mathbb{P}(L)$ (Section 6).

The linearized problem associated with (1.1) near $\mathbb{P}(L)$ is studied in Section 7 and its hyperbolicity in directions transversal to $T_{\mathbf{m}}\mathbb{P}(L)$ is verified there.

Based on these results, in Section 8 we construct the center manifold reduction for equation (1.1) in a small neighborhood of the multi-pulse manifold $\mathbb{P}(L)$. In particular, Theorems 1.1 and 1.2 are proven there.

The relations between hyperbolicity and stability in infinite dimensional pulse configurations, which are very important for the applications to space-time chaos, are investigated in Section 9.

The asymptotic expansions of the reduced equations on the manifold $\mathbb{P}(L)$ as $L \to \infty$ are studied in Section 10. In particular, explicit formulae for the leading term in these expansions are computed for a number of concrete equations of mathematical physics.

Finally, the application of the above theory to space-time chaos in the 1D Swift-Hohenberg equations is given in Section 11.

2. Assumptions and preliminaries

In this section, we formulate the basic assumptions to the equation considered, its pulse solution and the associated symmetry group and introduce some useful notations which are assumed to be valid throughout the paper. We start with describing the general structure of our basic nonlinear PDE.

2.1. The equation. In the whole physical space $\Omega = \mathbb{R}^n$ we consider the following parabolic system of PDEs:

$$\partial_t u + A_0 u + \Phi(u) = 0. \tag{2.1}$$

It is assumed that $u(t,x) := (u^1(t,x), \cdots, u^m(t,x))$ is an unknown vector-valued function, $A_0 : [W^{2l,2}(\mathbb{R}^n)]^m \to [L^2(\mathbb{R}^n)]^m$ is a positive, strongly elliptic differential operator of order $2l$ with constant coefficients:

$$(A_0 u)_i := \sum_{j=1}^{m} \sum_{|p|,|q|=0}^{l} \partial_x^p (a_{p,q,i,j} \partial_x^q u_j) \tag{2.2}$$

where $p, q \in \mathbb{Z}_+^n$ are multi-indices, $\partial_x^p := \partial_{x^1}^{p_1} \cdots \partial_{x^n}^{p_n}$ and $a_{p,q,i,j} \in \mathbb{R}$ and the nonlinear term $\Phi(u)$ has the form

$$\Phi(u) := \bar{\Phi}(u, D_x^1 u, \cdots, D_x^{2l-1} u) \tag{2.3}$$

where D_x^s is collection of all derivatives ∂_x^p, $p \in \mathbb{Z}_+^n$, $|p| = s$ and $\bar{\Phi}$ is a smooth function. Moreover, we assume that

$$\Phi(0) = D_u \Phi(0) = 0. \tag{2.4}$$

We recall that the strong ellipticity of the operator A_0 means that its principal symbol

$$\mathcal{A}(\xi)_{i,j} := (-1)^l \sum_{|p|=|q|=l}^{l} \xi^{p+q} a_{p,q,i,j} \tag{2.5}$$

(where as usual $\xi^p := \xi_1^{p_1} \cdots \xi^{p_n}$) satisfies

$$\mathcal{A}(\xi) v.v \geq c |\xi|^{2l} |v|^2, \quad \xi \in \mathbb{R}^n, \quad v \in \mathbb{R}^m \tag{2.6}$$

for some positive $c > 0$, see e.g. [**Ama95**], and the positivity means that the spectrum of A_0 has strictly positive real part

$$\operatorname{Re} \sigma(A_0, [L^2(\mathbb{R}^n)]^m) > C_1 > 0. \tag{2.7}$$

REMARK 2.1. It is well-known (see e.g. [**Ama95**]), that under the above assumptions the elliptic differential operator A_0 defines an isomorphism between the Sobolev spaces $[W^{2l,p}(\mathbb{R}^n)]^m$ and $[L^p(\mathbb{R}^n)]^m$ for all $1 < p < \infty$ and generates an analytic semigroup in that spaces. In particular, the parabolic operator $\partial_t + A_0$ is also an isomorphism between the Sobolev-Slobodetskij space $[W^{(1,2l),p}(\mathbb{R}^{n+1})]^m$ and $[L^p(\mathbb{R}^{n+1})]^m$. Actually, we impose the strong ellipticity assumption on A_0 *only* in order to guarantee these properties. Thus, all of the results formulated below remains valid for operators A_0 which satisfy the above isomorphism properties.

We are now going to describe the symmetry group of the introduced basic equation.

2.2. The symmetry group. First of all, we recall that equation (2.1) always possesses a group of translations $\{T_\xi, \xi \in \mathbb{R}^n\}$, $(T_\xi u)(x) := u(x - \xi)$. Except of that group, this equation very often possesses a (compact) symmetry group of "rotations" (e.g., $SO(n)$ or $SO(n) \times SO(m)$ or subgroups thereof). That is the reason why we assume that the symmetry group G of equation (2.1) is a linear representation of a finite-dimensional Lie group G_0,

$$G := \{\mathcal{T}_\Gamma, \ \Gamma \in G_0\} \subset \mathcal{L}([C^\infty(\mathbb{R}^n)]^m, [C^\infty(\mathbb{R}^n)]^m), \ \ \mathcal{T}_{\Gamma_1} \circ \mathcal{T}_{\Gamma_2} = \mathcal{T}_{\Gamma_1 \circ \Gamma_2} \tag{2.8}$$

$\dim G_0 = k \geq n$, which is a skew product of $\mathbb{R}^n$ and some $(k-n)$-dimensional compact Lie group $\bar{G}_0$, i.e., every element $\Gamma \in G_0$ has the form $\Gamma = (\xi, \gamma)$, $\xi \in \mathbb{R}^n$, $\gamma \in \bar{G}_0$ and elements $\xi \in \mathbb{R}^n$ generate the subgroup of translations in G:

$$\mathcal{T}_{(\xi,\gamma)} = T_\xi \circ \mathcal{T}_{(0,\gamma)}. \tag{2.9}$$

For simplicity, we assume that G_0 is realized as a closed subgroup of the matrix group $GL(\mathbb{R}^N) \subset \mathbb{R}^{N\times N}$ for some large N and, consequently, $\bar{G}_0 \subset GL(\mathbb{R}^N)$ is a compact subgroup of it. Furthermore, we can define a metric on the group G_0 via

$$I(\Gamma_1, \Gamma_2) = \|\Gamma_1 - \Gamma_2\| := \|\gamma_1 - \gamma_2\|_{\bar{G}_0} + \|\xi_1 - \xi_2\|_{\mathbb{R}^n}$$

where $\|\cdot\|_{\bar{G}_0}$ is some norm on the space of matrices $M(\mathbb{R}^N) = \mathbb{R}^{N\times N}$, which is left invariant with respect to $\bar{G}_0 \subset GL(\mathbb{R}^N)$ (which exists since $\bar{G}_0$ is compact) and $\|\cdot\|_{\mathbb{R}^n}$ is a standard norm in $\mathbb{R}^n$. Then, the metric thus defined on G_0 will be also left invariant:

$$I(\Gamma\circ\Gamma_1, \Gamma\circ\Gamma_2) = I(\Gamma_1, \Gamma_2).$$

Moreover, we assume that the action $\Gamma \to \mathcal{T}_\Gamma$ is smooth with respect to $\Gamma \in G$ at every $\phi(x)$ satisfying

$$|\partial_x^p \phi(x)| \le C_p\, \mathrm{e}^{-\alpha|x|} \tag{2.10}$$

(where the constant C_p depends on $p \in \mathbb{Z}_+^m$, but is independent of $x \in \mathbb{R}^n$) and the following inequalities hold:

$$\begin{aligned} \|\partial_\Gamma^p(\mathcal{T}_\Gamma\phi)(x)\| \cdot (1 + |x-\xi|^{|p|})^{-1} &+ \\ + |\partial_x^p(\mathcal{T}_\Gamma\phi)(x)| \le C_p'\, \mathrm{e}^{-\alpha|x-\xi|}, \quad &\forall x \in \mathbb{R}^n, \quad \Gamma = (\xi,\gamma) \in G_0 \end{aligned} \tag{2.11}$$

where the norms of ∂_Γ^p-derivatives are computed using the metric associated with the above embedding $G_0 \subset GL(\mathbb{R}^N)$ and the constant C_p' depends on p through the constant C_p from (2.10), but is independent of Γ and x.

Furthermore, in order to avoid the technicalities, we make two additional assumptions on the symmetry group G. The first is that it can be extended to $\mathcal{L}([W^{l,p}(\mathbb{R}^n)]^m, [W^{l,p}(\mathbb{R}^n)]^m)$ for all $1 \le p \le \infty$ and $l \in \mathbb{Z}_+$ and

$$\|\mathcal{T}\|_{\mathcal{L}([W^{l,p}(\mathbb{R}^n)]^m, [W^{l,p}(\mathbb{R}^n)]^m)} \le C_{l,p}, \quad \forall \mathcal{T} \in G \tag{2.12}$$

where the constant C depends on p and l, but is independent of $\mathcal{T} \in G$. And the second one is that G acts by isometries in $[L^2(\mathbb{R}^n)]^m$, i.e

$$(\mathcal{T}u, \mathcal{T}v) = (u, v) \ \text{ for all } \ \mathcal{T} \in G. \tag{2.13}$$

It is also worth to mention that, due to (2.4), G is a symmetry group not only for whole equation (2.1), but also for the operators A_0 and Φ separately, i.e.

$$\mathcal{T}\circ A_0 = A_0 \circ \mathcal{T}, \quad \mathcal{T}\circ\Phi = \Phi\circ\mathcal{T}, \quad \forall \mathcal{T} \in G. \tag{2.14}$$

All of the assumptions of this subsection are obviously satisfied if $\bar{G}_0$ is a subgroup of $SO(n) \times SO(m)$ with the standard action on $[C^\infty(\mathbb{R}^n)]^m$. In all applications of the general theory considered in the paper the group $\bar{G}_0$ will be such a subgroup. Consequently, it will be not necessary to verify the above conditions in the concrete examples considered below (see Section 10).

Finally, we formulate the assumptions on the localized pulse solution.

2.3. The pulse. We assume that equation (2.1) possesses a smooth pulse equilibrium $V(x)$ which satisfies the following estimate:

$$|\partial_x^p V(x)| \le C_p\, \mathrm{e}^{-\alpha|x|} \tag{2.15}$$

where the constant C_p depends on $p \in \mathbb{Z}_+^m$, but is independent of $x \in \mathbb{R}^n$.

Let us consider the linearization of (2.1) on the pulse solution V which gives the following linear elliptic operator:

$$\mathcal{L}_V := A_0 + D_u\Phi(V). \tag{2.16}$$

We recall that, due to our assumptions on A_0 and Φ, the operator A_0 is invertible and $D_u\Phi(V)$ is relatively compact with respect to A_0. Thus, (2.16) is a Fredholm operator of index zero and, consequently, has a finite-dimensional kernel and co-kernel. On the other hand, since equation (2.1) is invariant with respect to $\mathcal{T}_\Gamma$, then

$$\phi^i := L_{v^i}(\mathcal{T}_\Gamma)V\big|_{\Gamma=\Gamma_e} \in \ker \mathcal{L}_V, \quad i = 1, \cdots, k \tag{2.17}$$

where $\Gamma_e := (0, e)$ is a unit element of the group G_0, $\{v^i\}_{i=1}^k$ is some fixed basis in the Lie algebra $\mathcal{G}$ associated with the group G_0 and L_{v^i} is a differentiation along v^i. To be more precise, let $\exp : \mathcal{G} \to G_0$ be the associated exponential map. Since G_0 is realized as a subgroup of the matrix group $GL(\mathbb{R}^N)$, the exponential map $\exp(\cdot)$ coincides with the standard matrix exponent:

$$\exp(\mathrm{g}) := \sum_{l=0}^{\infty} \frac{\mathrm{g}^l}{l!}, \quad \mathrm{g} \in \mathcal{G} \subset M(\mathbb{R}^N). \tag{2.18}$$

Then, the basis $\{v^i\}_{i=1}^k$ generates k one-parametrical subgroups $\{\exp(sv^i),\ s \in \mathbb{R}\}$ in G_0 and the functions ϕ^i can be defined as follows:

$$\phi^i = \frac{d}{ds}\mathcal{T}_{\exp(sv^i)}V\big|_{s=0}.$$

Our next assumption is that the eigenvectors ϕ^i are linearly independent

$$\det\{(\phi^i, \phi^j)\} \neq 0 \tag{2.19}$$

and generate the whole kernel $\ker \mathcal{L}_V$:

$$\ker \mathcal{L}_V = \operatorname{span}\{\phi^i,\ i = 1, \cdots, k\}, \quad \dim \ker \mathcal{L}_V = k.$$

Moreover, we assume that the operator $\mathcal{L}_V$ does not have any Jordan cells at zero eigenvalue (i.e. that its geometric multiplicity is equal to the algebraic one) and the rest of the spectrum of $\mathcal{L}_V$ is separated from the imaginary axis:

$$\sigma(\mathcal{L}_V, L^2(\mathbb{R}^n)) \cap \{\lambda \in \mathbb{C},\ \operatorname{Re}\lambda \in [-\delta, \delta]\} = \{0\} \tag{2.20}$$

for some positive δ.

Note also that, according to (2.11) eigenfunctions $\phi^i(x)$ are smooth and exponentially decaying (i.e. satisfy (2.10)). Furthermore, since $\mathcal{L}_V$ is Fredholm of index zero, then the adjoint operator

$$\mathcal{L}_V^* := A_0^* + [D_u\Phi(V)]^* \tag{2.21}$$

also has k-dimensional kernel

$$\ker \mathcal{L}_V^* := \operatorname{span}\{\psi^i,\ \ i = 1, \cdots, k\}. \tag{2.22}$$

We also note that, according to our assumptions on Φ and V, all of the coefficients of linear differential operators $D_u\Phi(V)$ and $[D_u\Phi(V)]^*$ satisfy estimate (2.10) and, consequently, using the standard elliptic regularity estimates in weighted spaces (see e.g. [**Tri78**] or the next section), it is not difficult to verify that the eigenfunctions ψ^i are also smooth and exponentially decaying (i.e., satisfy (2.10)). We assume that the eigenfunctions $\psi^i \in [L^2(\mathbb{R}^n)]^m$ of the adjoint operator are normalized in such way that

$$(\phi^i, \psi^j) = \delta_{ij}. \tag{2.23}$$

Such a normalization exists since, due to our assumptions, the operator $\mathcal{L}_V$ does not have any Jordan cells at zero eigenvalue. It is also worth to recall that, according to Fredholm property, equation

$$\mathcal{L}_V v = h, \quad (v, \psi^i) = 0, \quad i = 1, \cdots, k$$

is uniquely solvable if and only if $(h, \psi^i) = 0$, $i = 1, \cdots, k$, then the following estimate holds:

$$\|v\|_{[W^{2l,p}(\mathbb{R}^n)]^m} \le C_p \|h\|_{[L^p(\mathbb{R}^n)]^m} \tag{2.24}$$

where the constant C_p depends on $1 < p < \infty$, but is independent of h (and the analogous result holds also for the adjoint operator).

We say that the initial pulse $V(x)$ is *spectrally stable* if the following stronger analog of (2.20) holds:

$$\sigma(\mathcal{L}_V, L^2(\mathbb{R}^n)) \cap \{\lambda \in \mathbb{C},\ \operatorname{Re}\lambda \in [-\delta, \infty]\} = \{0\}. \tag{2.25}$$

In other words, the pulse $V(x)$ is spectrally stable, if the spectrum of $\mathcal{L}_V$ consists of a zero eigenvalue of finite multiplicity and the rest of the spectrum belongs to the left halfplane.

And, finally, we assume for simplicity that the stabilizer of V is trivial, see Remark 2.5 for some discussion of the general case:

$$\operatorname{St}_G(V) := \{\Gamma \in \Gamma_0,\ \mathcal{T}_\Gamma V = V\} = \{\Gamma_e\}. \tag{2.26}$$

We recall that together with the principal pulse solution $V(x)$, we also have a whole family of pulse solutions $V_\Gamma(x) := (\mathcal{T}_\Gamma V)(x)$ parameterized by the elements of the symmetry group G_0 and, consequently, equation (2.1) possesses the following one-pulse manifold of equilibria:

$$\mathbb{P}_1 := \{\mathcal{T}_\Gamma V,\ \Gamma \in G_0\}. \tag{2.27}$$

We conclude the section by formulating the corollaries of the above assumptions on the structure of this manifold which are of fundamental significance for what follows.

2.4. The one-pulse invariant manifold: uniform normal hyperbolicity. We first recall that, due to (2.11), all the pulses V_Γ satisfy the following shifted analog of estimate (2.10):

$$|\partial_x^p V_\Gamma(x)| \le C_p e^{-\alpha|x-\xi|}, \quad \Gamma := (\xi, \gamma). \tag{2.28}$$

where the constant C_p depends only on p. Let us now consider the linearized operator $\mathcal{L}_\Gamma$ on the pulse V_Γ

$$\mathcal{L}_\Gamma := A_0 + D_u\Phi(V_\Gamma) := A_0 + F_\Gamma. \tag{2.29}$$

Then, all of the coefficients of linear differential operator $F_\Gamma := D_u\Phi(V_\Gamma)$ also satisfy (2.28). Moreover, due to the commutation relations (2.14),

$$\mathcal{L}_\Gamma = \mathcal{T}_\Gamma \circ \mathcal{L}_V \circ (\mathcal{T}_\Gamma)^{-1}, \quad \mathcal{L}_\Gamma^* = \mathcal{T}_\Gamma \circ \mathcal{L}_V^* \circ (\mathcal{T}_\Gamma)^{-1} \tag{2.30}$$

(here we have used our simplifying assumption that $\mathcal{T}_\Gamma$ are isometries in L^2 and, consequently, $\mathcal{T}_\Gamma^* = (\mathcal{T}_\Gamma)^{-1}$). Thus, the functions $\phi_\Gamma^i := \mathcal{T}_\Gamma\phi^i$ and $\psi_\Gamma^i := \mathcal{T}_\Gamma\psi^i$, $i = 1, \cdots, k$ generate the kernel and co-kernel of the operator $\mathcal{L}_\Gamma$ respectively:

$$\ker \mathcal{L}_\Gamma = \operatorname{span}\{\phi_\Gamma^i,\ \ i = 1, \cdots, k\}, \ \ \ker \mathcal{L}_\Gamma^* = \operatorname{span}\{\psi_\Gamma^i,\ \ i = 1, \cdots, k\} \tag{2.31}$$

and satisfy the shifted analog of (2.10):

$$|\partial_x^p \phi_\Gamma^i(x)| + |\partial_x^p \psi_\Gamma^i(x)| \le C_p\, e^{-\alpha|x-\xi|}\,. \tag{2.32}$$

Moreover, due to (2.12) the equation

$$\mathcal{L}_\Gamma v = h, \ \ (v, \psi_\Gamma^i) = 0 \tag{2.33}$$

is uniquely solvable if and only if $(h, \psi_\Gamma^i) = 0$, then the estimate

$$\|v\|_{[W^{2l,p}(\mathbb{R}^n)]^m} \le C_p \|h\|_{[L^p(\mathbb{R}^n)]^m} \tag{2.34}$$

holds *uniformly* with respect to $\Gamma \in G_0$ and, since $\mathcal{T}_\Gamma$ are isometries

$$\det\{(\phi_\Gamma^i, \phi_\Gamma^j)\} = \det\{(\phi^i, \phi^j)\} \ne 0, \ \ \det\{(\psi_\Gamma^i, \psi_\Gamma^j)\} = \det\{(\psi^i, \psi^j)\} \ne 0, \\ (\phi_\Gamma^i, \psi_\Gamma^j) = (\phi^i, \psi^j) = \delta_{ij}. \tag{2.35}$$

Thus, the invariant manifold $\mathbb{P}_1$ is indeed uniformly normally hyperbolic with respect to equation (2.1).

We are now going to study the local structure of the manifold $\mathbb{P}_1$. To this end, we consider the local coordinates $\Gamma = \Gamma(\alpha)$, $\alpha = (\alpha^1, \cdots, \alpha^k) \in \mathbb{R}^k$ near some point $\Gamma_0 \in G_0$. Then, differentiating the equation $A_0 V_{\Gamma(\alpha)} + \Phi(V_{\Gamma(\alpha)}) = 0$ with respect to α^i, we derive that

$$\partial_{\alpha^i} V_{\Gamma(\alpha)} \in \ker \mathcal{L}_{\Gamma(\alpha)} \tag{2.36}$$

and, consequently, there exists a matrix $\Pi_{ij}(\alpha) \in \mathcal{L}(\mathbb{R}^k, \mathbb{R}^k)$ such that

$$\partial_{\alpha^i} V_{\Gamma(\alpha)} = \sum_{j=1}^k \Pi_{ij}(\alpha)\phi^j_{\Gamma(\alpha)}. \tag{2.37}$$

The next proposition shows that $\mathbb{P}_1$ is a manifold globally diffeomorphic to G_0 and that there exists a convenient coordinate atlas on $\mathbb{P}_1$ such in all local coordinates of that atlas the matrices Π_{ij} are uniformly bounded and invertible.

PROPOSITION 2.2. *Let the above assumptions hold. Then there exists a positive constant C_0 such that*

$$C_0^{-1} \min\{1, \|\Gamma_1 - \Gamma_2\|\} \le \|V_{\Gamma_1} - V_{\Gamma_2}\|_{[L^2(R^n)]^m} \le C_0 \min\{1, \|\Gamma_1 - \Gamma_2\|\} \tag{2.38}$$

for all $\Gamma_1, \Gamma_2 \in G_0$ and the space L^2 in (2.38) can be replaced by any Sobolev space $W^{l,p}(\mathbb{R}^n)$ with $l \ge 0$ and $p \in [1, \infty]$.

Moreover, there exists a smooth coordinate atlas on $\mathbb{P}_1$ such that, for every local coordinates of this atlas, we have

$$\|\partial_\alpha^p \Pi(\alpha)\|_{\mathcal{L}(\mathbb{R}^k,\mathbb{R}^k)} + \|\Pi^{-1}(\alpha)\|_{\mathcal{L}(\mathbb{R}^k,\mathbb{R}^k)} \le C_p, \tag{2.39}$$

where the constant C_p depends only on p and is independent of α and on the concrete choice of the local coordinates.

Furthermore, there exists positive r_0 such that, for every $\Gamma_0 \in G_0$ there exist local coordinates α near V_{Γ_0} belonging to the above atlas such that the set

$$\mathcal{O}_{r_0}(V_{\Gamma_0}) := \{V_\Gamma,\ \|\Gamma - \Gamma_0\| \le r_0\}$$

belongs to the same coordinate neighborhood and

$$(2.40)\quad C_0^{-1}\|\alpha_1-\alpha_2\|_{\mathbb{R}^k}\le\|V_{\Gamma(\alpha_1)}-V_{\Gamma(\alpha_2)}\|_{[L^2(\mathbb{R}^n)]^m}\le \\ \le C_0\|\alpha_1-\alpha_2\|_{\mathbb{R}^k},\quad \Gamma(\alpha_1),\Gamma(\alpha_2)\in\mathcal{O}_{r_0}(V_{\Gamma_0})$$

where the constant C_0 *is independent of* Γ_0, α_1 *and* α_2 *(and, analogously to* (2.38)*, the* L^2*-norm can be replaced by any Sobolev's norm* $W^{l,p}$*).*

PROOF. Indeed, since the map $\Gamma\to V_\Gamma$ is smooth and the vectors $L_{v^i}V_\Gamma\big|_{\Gamma=\Gamma_e}$, $i=1,\cdots,k$ are linear independent, then this map is a local diffeomorphism between G_0 and $\mathbb{P}_1$ near $\Gamma=\Gamma_e$. Thus, there exists a neighborhood $\mathcal{O}_{r_0}(\Gamma_e)$ of the unitary element such that

$$(2.41)\qquad C_0^{-1}\|\Gamma_1-\Gamma_2\|\le\|V_{\Gamma_1}-V_{\Gamma_2}\|_{L^2(\mathbb{R}^n)}\le C_0\|\Gamma_1-\Gamma_2\|$$

for all $\Gamma_1,\Gamma_2\in\mathcal{O}_{r_0}(\Gamma_e)$. Moreover, since all norms on a finite-dimensional space are equivalent, the L^2-norm in the middle part of (2.41) can be replaced by any Sobolev norm $W^{l,p}$.

Furthermore, it is well-known that the exponential map

$$\Gamma(\alpha):=\exp(\alpha\cdot v),\quad \alpha\cdot v:=\sum_{i=1}^k\alpha^iv^i,$$

where $\alpha\in\mathbb{R}^k$ and $\{v^i\}_{i=1}^k$ is a basis in the Lie algebra $\mathcal{G}$, is a local diffeomorphism near $\alpha=0$ and defines, thus, the local coordinates on G_0 near the unit element $\Gamma=\Gamma_e$. Lifting the constructed local coordinates from G_0 to $\mathbb{P}_1$, we obtain the local coordinates $\alpha\to V_{\Gamma(\alpha)}$ on $\mathbb{P}_1$ near the initial pulse $V=V_{\Gamma_e}$. Obviously, the coordinates thus defined satisfy (2.40) for every $\Gamma_1,\Gamma_2\in\mathcal{O}_{r_0}(\Gamma_e)$. Finally, since $\{\phi^i_{\Gamma(\alpha)}\}_{i=1}^k$ and $\{\partial_{\alpha^i}V_{\Gamma(\alpha)}\}_{i=1}$ are bases in $\ker\mathcal{L}_{\Gamma(\alpha)}$, the transfer matrix $\Pi_{ij}(\alpha)$ is invertible and satisfies (2.39) uniformly with respect to $\Gamma\in\mathcal{O}_{r_0}(\Gamma_e)$.

Thus, we have verified all the assumptions of the proposition *locally* near the unitary point Γ_e. In order to extend these local results from $\mathcal{O}_{r_0}(\Gamma_e)$ to $\mathcal{O}_{r_0}(\Gamma_0)$, for every $\Gamma_0\in G_0$, it is sufficient to "shift" them by the group action operator $\mathcal{T}_{\Gamma_0}$. Indeed, the local coordinates $\alpha\to\Gamma(\alpha)$ near arbitrary $\Gamma_0\in G_0$ can be defined via

$$\Gamma(\alpha):=\Gamma_0\circ\exp(\alpha\cdot v),\quad \alpha\cdot v:=\sum_{i=1}^k\alpha^iv^i,$$

where $\alpha\in\mathbb{R}^k$. Then, due to assumption (2.12) and the proved estimates near $\Gamma=\Gamma_e$, the coordinate atlas thus defined satisfies all the assertions of the proposition. In particular, since

$$\phi^i_{\Gamma(\alpha)}=\mathcal{T}_{\Gamma_0}\phi^i_{\exp(\alpha\cdot v)},\quad \partial_{\alpha^i}V_{\Gamma(\alpha)}=\mathcal{T}_{\Gamma_0}\partial_{\alpha^i}V_{\exp(\alpha\cdot v)},$$

the transfer matrix $\Pi(\alpha)$ is independent of Γ_0 and (2.39) holds.

Moreover, the analogous shift arguments, together with the assumption that the metric $\|\Gamma_1-\Gamma_2\|$ on Γ_0 is left invariant, show that estimate (2.41) holds for every $\Gamma_1,\Gamma_2\in G_0$ such that $\|\Gamma_1-\Gamma_2\|\le r_0$. Thus, it only remains to verify (2.38) in the case $\|\Gamma_1-\Gamma_2\|\ge r_0$.

To this end, we note that the right-hand side of this inequality is obvious since $\mathbb{P}_1$ is globally bounded in $[L^2(\mathbb{R}^n)]^m$ (and (2.41) holds for every Γ_1 and Γ_2 which are sufficiently close to each other), so we only need to verify the left one. Assume that this estimate is wrong. Then, there exists two sequences Γ_i^1 and Γ_i^2 such that

$$\|\Gamma_i^1 - \Gamma_i^2\| \ge r_0, \quad \|V_{\Gamma_i^1} - V_{\Gamma_i^2}\|_{[L^2(\mathbb{R}^n)]^m} \le 1/i. \tag{2.42}$$

Moreover, due to the above translation trick, we can assume without loss of generality that $\Gamma_i^1 = \Gamma_e$ and $\Gamma_i^2 = \Gamma_i = (\xi_i, \gamma_i)$. We also note that, the pulse V_{Γ_i} is "localized" near $x = \xi_i$ (see (2.28)) and, therefore, (2.42) implies that ξ_i are uniformly bounded: $|\xi_i| \le R$, $i \in \mathbb{N}$. Taking now into account that G_0 is a product of $\mathbb{R}^n$ and a compact group, we may assume that $\Gamma_i \to \Gamma$ in G_0 as $i \to \infty$. Finally, passing to the limit in (2.42), we infer

$$\Gamma_e \ne \Gamma \in \mathrm{St}_G(V)$$

which contradicts assumption (2.26) on the triviality of this stabilizer. Proposition (2.2) is proven. □

REMARK 2.3. If it is known, in addition, that the Lie group G_0 is commutative, the above defined natural local coordinates $\Gamma(\alpha)$ near $\Gamma = \Gamma_e$ have the form

$$\Gamma(\alpha) = \exp(\alpha^1 v^1) \cdot \dots \cdot \exp(\alpha^k v^k)$$

where $\{v^i\}_{i=1}^k$ is the above fixed basis in the Lie algebra. Consequently, in these coordinates the transfer matric Π is identical

$$\Pi(s) \equiv \mathrm{Id}\,.$$

However, for general non-commutative Lie groups we do not have such canonical coordinates and the only thing which we can assume is

$$\Pi(\Gamma_e) = \mathrm{Id}$$

(just taking the coordinate axis at $\Gamma = \Gamma_e$ along the vectors v_i). Translating this local coordinates to arbitrary point $\Gamma = \Gamma_0$, we see that there exists a coordinate system from the atlas described in the previous proposition *centered* at $\Gamma = \Gamma_0$ such that $\Pi(\Gamma_0) = \mathrm{Id}$.

We conclude by some kind of spatially localized version of estimate (2.38) which will be useful for what follows.

PROPOSITION 2.4. *Let the above assumptions hold. Then, there exists positive R_0 such that, for every $\Gamma = (\xi, \gamma)$ and $\Gamma' = (\xi', \gamma')$ belonging to G_0, we have*

$$(C_0')^{-1} \min\{1, \|\Gamma - \Gamma'\|\} \le \|V_\Gamma - V_{\Gamma'}\|_{L^2(B_\xi^{R_0})} \le C_0' \min\{1, \|\Gamma - \Gamma'\|\}. \tag{2.43}$$

where $B_{x_0}^R$ denotes the ball of radius R in the space $\mathbb{R}^n$ centered at x_0, the constant C_0 is independent of $\Gamma, \Gamma' \in G_0$ and the L^2-norm can be replaced by any Sobolev's norm $W^{l,p}$.

PROOF. We first note that, due to (2.38), it is sufficient to verify only the left inequality of (2.43). Moreover, due to translation invariance, we may assume without loss of generality that $\xi = 0$. Furthermore, using the assumption $\mathrm{St}_G(V) = \{\Gamma_e\}$ analogously to the previous proposition, we reduce the proof to the case where $\|\Gamma - \Gamma'\| \le r_0$ and, consequently, Γ and Γ' belong to the same coordinate neighborhood.

Using now assumption (2.11), we infer

$$|V_\Gamma(x) - V_{\Gamma'}(x)| \le C\|\Gamma - \Gamma'\| \int_0^1 |\partial_\Gamma V_{\Gamma_s}(x)|\, ds \le C_r \|\Gamma - \Gamma'\|\, e^{-\alpha|x|/2}$$

where Γ_s is a segment (in local coordinates) in G_0 connecting Γ and Γ'. The last formula, in turns, gives an estimate

$$\|V_\Gamma - V_{\Gamma'}\|_{L^2(\mathbb{R}^n\setminus B_0^{R_0})} \le C_r\|\Gamma - \Gamma'\|/R_0 \tag{2.44}$$

Thus, using the left estimate of (2.38) and (2.44), we infer

$$\|V_\Gamma - V_{\Gamma'}\|^2_{L^2(B_0^{R_0})} = \|V_\Gamma - V_{\Gamma'}\|^2_{L^2(\mathbb{R}^n)} - \\ - \|V_\Gamma - V_{\Gamma'}\|^2_{L^2(\mathbb{R}^n\setminus B_0^{R_0})} \ge (C_0^{-2} - C_r^2/R_0^2)\|\Gamma - \Gamma'\|^2$$

which implies (2.43) if R_0 is large enough. Proposition 2.4 is proven. □

REMARK 2.5. The properties of the normally hyperbolic one-pulse invariant manifold $\mathbb{P}_1$ collected in the previous subsection play a fundamental role throughout of the paper. In contrast to this, the fact that this manifold is generated from a single pulse by translation via the symmetry group is not essential and will be nowhere essentially used below. In a fact, we have introduced the symmetry formalism only in order to simplify the verification of the above uniform normal hyperbolicity and to avoid the technicalities although the most part of the results of the paper remain true for general one-pulse invariant manifolds satisfying the properties of the previous subsection.

In particular, if the initial pulse has a non-trivial stabilizer $\mathrm{St}_G(V)$ and this stabilizer is not a *normal* subgroup of G, then the situation cannot be formally reduced to the case considered (since the quotient $G/\,\mathrm{St}_G(V)$ is not a group). However, the symmetry group G still acts transitively on $\mathbb{P}_1$ and, consequently, the *uniform* normal hyperbolicity of $\mathbb{P}_1$ follows as before from the normal hyperbolicity of a single initial pulse V. Thus, the theory developed below can be extended with minor changes to the general case of non-trivial stabilizer. The only difference is related with the fact that the local coordinates on the quotient $G/\,\mathrm{St}_G(V)$ have slightly more complicated structure and this, in turns, can lead to more complicated form of the asymptotic expansions of pulse interaction equations in local coordinates, see Section 10.

3. Weighted Sobolev spaces and regularity of solutions

In this section we recall some basic facts concerning weighted Sobolev spaces with weights of exponential growth rate and regularity of solutions of elliptic and parabolic equations in that spaces which are of fundamental significance for the next sections. We start by introducing the class admissible of weight functions.

DEFINITION 3.1. A function $\theta \in C(\mathbb{R}^N)$ is a weight function of exponential growth rate $\varepsilon \ge 0$ if $\theta(z) > 0$ for all $z \in \mathbb{R}^N$ and

$$\theta(z_1 + z_2) \le C_\theta \theta(z_1)\,\mathrm{e}^{\varepsilon|z_2|}, \tag{3.1}$$

for all $z_1, z_2 \in \mathbb{R}^N$ and some positive constant C_θ depending only on θ.

We will mainly use the following basic weight functions of exponential growth rate:

$$\bar\theta_{\varepsilon,z_0}(z) := \mathrm{e}^{-\varepsilon|z-z_0|} \tag{3.2}$$

where $\varepsilon \in \mathbb{R}$ and $z_0 \in \mathbb{R}^N$ are parameters and their smooth analogs

$$\theta_{\varepsilon,z_0}(z) := \mathrm{e}^{-\varepsilon\sqrt{|z-z_0|^2+1}}\,. \tag{3.3}$$

Obviously all these functions have exponential growth rate $|\varepsilon|$ and the constant C_θ for these weights are independent of $z_0 \in \mathbb{R}^N$.

The next definition gives weighted Sobolev spaces associated with these weights.

DEFINITION 3.2. Let θ be a weight function of exponential growth rate. For every, $1 \le p \le \infty$, we define the spaces $L^p_\theta(\mathbb{R}^N)$ and $L^p_{b,\theta}(\mathbb{R}^N)$ by the following norms:

$$\begin{aligned}\|u\|_{L^p_\theta(\mathbb{R}^N)} &:= \left(\int_{z\in\mathbb{R}^N} \theta^p(z)|u(z)|^p\,\mathrm{d}z\right)^{1/p}\\ \|u\|_{L^p_{b,\theta}(\mathbb{R}^N)} &:= \sup_{z\in\mathbb{R}^N}\left\{\theta(z)\|u\|_{L^p(B^1_z)}\right\}\end{aligned} \tag{3.4}$$

respectively (here and below $B^R_{x_0}$ denotes the R-ball of the space $\mathbb{R}^N$ centered at x_0). Moreover, for every $l \in \mathbb{N}$, we can define the Sobolev spaces $W^{l,p}_\theta(\mathbb{R}^N)$ and $W^{l,p}_{b,\theta}(\mathbb{R}^N)$ as the space of distributions whose derivatives up to order l belong to $L^p_\theta(\mathbb{R}^N)$ and $L^p_{b,\theta}(\mathbb{R}^N)$ respectively. In particular, in the case $N = n+1$, $z = (t,x)$, we can define the anisotropic Sobolev spaces $W^{(l_1,l_2),p}_\theta(\mathbb{R}^{n+1})$ and $W^{(l_1,l_2),p}_{b,\theta}(\mathbb{R}^{n+1})$ as the space of distributions whose t-derivatives up to order l_1 and x-derivatives up to order l_2 belong to $L^p_\theta(\mathbb{R}^{n+1})$ and $L^p_{b,\theta}(\mathbb{R}^{n+1})$ respectively. We will usually use these anisotropic spaces with exponents $(1, 2l)$ which are the usual spaces for the study of parabolic equations associated with the elliptic operators of order $2l$.

We collect in the next proposition the important estimates related with the above weighted spaces whose proofs and more detailed exposition can be found, e.g. in [**EfZ01, Zel03b**].

PROPOSITION 3.3. *Let $u \in L^p_\theta(\mathbb{R}^N)$ where θ is a weight function of exponential growth rate ε. Then,*

$$C_R(\theta)^{-1}\|u\|_{L^p_\theta(\mathbb{R}^N)} \le \left(\int_{z\in\mathbb{R}^N} \theta^p(z)\|u\|^p_{L^p(B^R_z)}\,\mathrm{d}z\right)^{1/p} \le C_R(\theta)\|u\|_{L^p_\theta(\mathbb{R}^N)} \tag{3.5}$$

where the constant $C_R(\theta)$ depends only on N, $R > 0$, ε and the constant C_θ and is independent of u, p and the concrete form of the weight function θ. Moreover, for every $\alpha > \varepsilon$, we have

$$\begin{aligned}&C_\alpha(\theta)^{-1}\|u\|_{L^p_\theta(\mathbb{R}^N)} \le\\ &\le \left(\int_{z_0\in\mathbb{R}^N} \theta^p(z_0)\int_{z\in\mathbb{R}^N} \mathrm{e}^{-p\alpha|z-z_0|}\,\|u\|^p_{L^p(B^1_z)}\,\mathrm{d}z\,\mathrm{d}z_0\right)^{1/p} \le C_\alpha(\theta)\|u\|_{L^p_\theta(\mathbb{R}^N)}\end{aligned} \tag{3.6}$$

where the constant $C_\alpha(\theta)$ depends only on N, α, ε and C_θ. Analogously, if $u \in L^p_{b,\theta}(\mathbb{R}^N)$ then

$$\begin{aligned}&C_\alpha^{-1}(\theta)\|u\|_{L^p_{b,\theta}(\mathbb{R}^N)} \le\\ &\le \sup_{z_0\in\mathbb{R}^N}\left\{\theta(z_0)\left(\int_{z\in\mathbb{R}^N} \mathrm{e}^{-p\alpha|z-z_0|}\,\|u\|^p_{L^p(B^1_z)}\,\mathrm{d}z\right)^{1/p}\right\} \le C_\alpha(\theta)\|u\|_{L^p_{b,\theta}(\mathbb{R}^N)}.\end{aligned} \tag{3.7}$$

Furthermore, estimates (3.5)–(3.7) remain valid if we replace thew spaces L^p by the Sobolev spaces $W^{l,p}$ and $W^{(l_1,l_2),p}$.

Estimates (3.5) are very useful in order to verify various embedding and interpolation estimates in weighted Sobolev spaces and estimates (3.6) and (3.7) allow us to obtain the regularity estimates for the solutions for arbitrary weights θ if the analogous estimates for special weights (3.2) or (3.3) are known.

We are ready now to recall the regularity estimates for the solutions of parabolic equations in weighted Sobolev spaces introduced before. We start with the following parabolic equation associated with the elliptic operator A_0 introduced in the previous section:

$$\partial_t v + A_0 v = h. \tag{3.8}$$

PROPOSITION 3.4. *Let the differential operator A_0 satisfies the assumptions of Section 2. Then, for every $1 < p < \infty$, there exists $\varepsilon_0 > 0$ depending only on A_0 and p such that, for every weight function $\theta = \theta(t,x)$ of exponential growth rate $\varepsilon \le \varepsilon_0$ and every $h \in [L^p_\theta(\mathbb{R}^n)]^m$ or $[L^p_{b,\theta}(\mathbb{R}^n)]^m$ there exists a unique solution v of equation (3.8) which belongs to $[W^{(1,2l),p}_\theta(\mathbb{R}^n)]^m$ (resp. $W^{(1,2l),p}_{b,\theta}(\mathbb{R}^n)]^m$) and the following estimates hold:*

$$\|v\|_{W^{(1,2l),p}_\theta(\mathbb{R}^n)} \le C\|h\|_{L^p_\theta(\mathbb{R}^n)}, \quad \text{resp.} \quad \|v\|_{W^{(1,2l),p}_{b,\theta}(\mathbb{R}^n)} \le C\|h\|_{L^p_{b,\theta}(\mathbb{R}^n)} \tag{3.9}$$

where the constant C depends on p and the constant C_θ from (3.1), but is independent of ε, h and the concrete form of the weight function θ.

PROOF. We first recall that, since A_0 is assumed to be uniformly elliptic and positively defined, then equation (3.8) is unique solvable for all $h \in L^p(\mathbb{R}^{n+1})$, $1 < p < \infty$ and the following estimate holds:

$$\|v\|_{W^{(1,2l),p}(\mathbb{R}^{n+1})} \le C_p\|h\|_{L^p(\mathbb{R}^{n+1})}, \tag{3.10}$$

see e.g. [**Ama95**]. Thus, estimate (3.9) is verified for the non-weighted case $\theta = 1$. In order to obtain its weighted analogies, we first consider the case of the special weights $\theta_{\varepsilon_0,z_0}(t,x)$, $z_0 := (t_0,x_0) \in \mathbb{R}^{n+1}$ introduced in (3.3). Indeed, let v be a solution of (2.8) and let $\tilde v(t,x) := v(t,x)\theta_{\varepsilon_0,z_0}(t,x)$. Then, it is not difficult to verify, using the obvious estimate

$$|\partial_z^p \theta_{\varepsilon_0,z_0}(t,x)| \le C_p\varepsilon_0\theta_{\varepsilon_0,z_0}(t,x), \quad (t,x) \in \mathbb{R}^{n+1} \tag{3.11}$$

(where the constant C_p depends on p, but is independent of (t,x), $\varepsilon_0 \le 1$ and z_0), that the function $\tilde v$ solves the following perturbed version of equation (3.8):

$$\partial_t \tilde v + A_0\tilde v = \varepsilon_0\tilde A_{\varepsilon_0}(t,x)\tilde v + \theta_{\varepsilon_0,z_0}h \tag{3.12}$$

where $\tilde A_0$ is the differential operator of order $(2l-1)$ with respect to x which is uniformly bounded with respect to ε_0 and z_0, i.e.

$$\|\tilde A_{\varepsilon_0}\tilde v(t)\|_{L^p(\mathbb{R}^n)} \le C\|\tilde v\|_{W^{2l-1,p}(\mathbb{R}^n)} \tag{3.13}$$

with the constant C independent of ε_0 and z_0. Applying now estimate (3.10) to equation (3.12) and using (3.13), we obtain that, for sufficiently small ε_0,

$$\|\tilde v\|_{W^{(1,2l),p}(\mathbb{R}^{n+1})} \le C_p'\|\theta_{\varepsilon_0,z_0}h\|_{L^p(\mathbb{R}^{n+1})}. \tag{3.14}$$

And, consequently, since the weights (3.3) are equivalent to (3.2), we have verified that

$$\|v\|^p_{W^{(1,2l),p}_{\tilde\theta_{\varepsilon_0,z_0}}(\mathbb{R}^{n+1})} \le C_p''\|h\|^p_{L^p_{\tilde\theta_{\varepsilon_0,z_0}}(\mathbb{R}^{n+1})} \tag{3.15}$$

where the constant C_p'' is independent of $z_0 \in \mathbb{R}^{n+1}$. Multiplying now estimate (3.15) by $\theta^p(z_0)$, integrating over $z_0 \in \mathbb{R}^{n+1}$ and using (3.6), we derive the first estimate of (3.9). The second one can be obtained analogously, replacing the integration over $z_0 \in \mathbb{R}^{n+1}$ by the supremum and using (3.7) instead of (3.6). Thus, estimates (3.9) are verified for every weight function θ of exponential growth rate $\varepsilon \le \varepsilon_0$. The existence of a solution follows in a standard way form these estimates and Proposition 3.4 is proven. □

We are going to obtain now the analogous estimates in weighted Sobolev spaces for the perturbed operators $\mathcal{L}_\Gamma := A_0 + F_\Gamma$, $\Gamma \in G_0$ introduced in (2.29) and the associated parabolic operators $\partial_t + \mathcal{L}_\Gamma$. In contrast to the operator A_0 these operators have k-dimensional kernels and that is the reason why we need to introduce the following projectors:

$$\mathbb{P}_\Gamma v := \sum_{i=1}^{k} (v, \psi_\Gamma^i)\phi_\Gamma^i, \tag{3.16}$$

where $\psi_\Gamma^i = \mathcal{T}_\Gamma \psi^i \in \ker \mathcal{L}_\Gamma^*$ and $\phi_\Gamma^i = \mathcal{T}_\Gamma \phi^i \in \ker \mathcal{L}_\Gamma$ are introduced in the previous section. Indeed, according to estimate (2.32) operators $\mathbb{P}_\Gamma$ are well defined for all $v \in L^2(\mathbb{R}^n)$ and uniformly bounded with respect to $\Gamma \in G_0$. Moreover, obviously,

$$\mathbb{P}_\Gamma^* v = \sum_{i=1}^{k} (v, \phi_\Gamma^i)\psi_\Gamma^i. \tag{3.17}$$

Instead of studying the equation $\mathcal{L}_\Gamma v = h$ and its parabolic analog (which is solvable only for h satisfying $\mathbb{P}_\Gamma h = 0$, see (2.33), (2.34)), it is much more convenient to consider the following modified equation:

$$\partial_t v + \mathcal{L}_\Gamma v + \mathbb{P}_\Gamma v = h \tag{3.18}$$

which is in a sense equivalent to the initial equation (see Remark 3.6 below), but is solvable for all $h \in L^p(\mathbb{R}^{n+1})$ as the following proposition shows.

PROPOSITION 3.5. *Let the operators $\mathcal{L}_\Gamma = A_0 + F_\Gamma$ satisfy all the assumptions imposed in Section 1. Then, equation (3.18) is uniquely solvable for every $h \in L^p(\mathbb{R}^{n+1})$ and the following estimate holds:*

$$\|v\|_{W^{(1,2l),p}(\mathbb{R}^{n+1})} \le C\|h\|_{L^p(\mathbb{R}^{n+1})} \tag{3.19}$$

where the constant C_p depends on p, but is independent of $\Gamma \in G_0$. Moreover, the analogous result holds also for the adjoint equation

$$-\partial_t w + \mathcal{L}_\Gamma^* w + \mathbb{P}_\Gamma^* w = h. \tag{3.20}$$

PROOF. We first recall that equations (3.20) are conjugate for different Γ by the transformations $\mathcal{T}_\Gamma$. Therefore, due to (2.12), it is sufficient to verify the solvability and estimate (3.19) for $\Gamma_e = (0, e)$ only. We now note that A_0 is a uniformly elliptic operator of order $2l$ and F_{Γ_e} is a differential operator of order $2l - 1$ whose coefficients satisfy estimate (2.32) and, consequently, decay exponentially as $|x| \to \infty$. Therefore, the term $F_{\Gamma_e} + \mathbb{P}_{\Gamma_e}$ is relatively compact with respect to A_0 and, consequently, the operator $\mathcal{L}_{\Gamma_e} + \mathbb{P}_{\Gamma_e}$ is a compact perturbation of a uniformly elliptic operator A_0. Thus, this operator also generates an analytic semigroup in $L^p(\mathbb{R}^n)$. Moreover, since $\mathbb{P}_{\Gamma_e} v \in \ker \mathcal{L}_{\Gamma_e}$, then the term $\mathbb{P}_{\Gamma_e} v$ does not change the component of the operator $\mathcal{L}_V$ (associated with the nonzero part of the spectrum) and the spectrum of the restriction of $\mathcal{L}_{\Gamma_e} + \mathbb{P}_{\Gamma_e}$ to the kernel $\ker \mathcal{L}_{\Gamma_e}$ obviously equals 1. Consequently,

$$\sigma(\mathcal{L}_{\Gamma_e} + \mathbb{P}_{\Gamma_e}) = \{1\} \cup (\sigma(\mathcal{L}_V)\backslash\{0\}) \tag{3.21}$$

and, due to assumption (2.20), there exists $\delta > 0$ with

$$\sigma(\mathcal{L}_{\Gamma_e} + \mathbb{P}_{\Gamma_e}) \cap \{\lambda \in \mathbb{C},\ \operatorname{Re}\lambda \in [-\delta, \delta]\} = \varnothing \tag{3.22}$$

and, consequently, operator $\mathcal{L}_{\Gamma_e} + \mathbb{P}_{\Gamma_e}$ possesses an exponential dichotomy in the space $[L^p(\mathbb{R}^n)]^m$, $1 < p < \infty$, which, in turns, implies the unique solvability of (3.18) and estimate (3.19), see e.g. [**Hen81**]. The adjoint equation (3.20) can be treated analogously. Proposition 3.5 is proven. □

REMARK 3.6. Let us multiply equation (3.18) scalarly by ψ^i_Γ. Then, taking into account the orthogonality conditions, we obtain the following system of ODE for determining $(v(t), \psi^i_\Gamma)$:

$$\frac{d}{dt}(v(t), \psi^i_\Gamma) + (v(t), \psi^i_\Gamma) = (h(t), \psi^i_\Gamma), \quad i = 1, \cdots, k \tag{3.23}$$

which shows that if $\mathbb{P}_\Gamma h(t) \equiv 0$, then $\mathbb{P}_\Gamma v(t) \equiv 0$ (if $v(t)$ does not grow very fast as $t \to \infty$) and consequently v solves equation without the artificial term $\mathbb{P}_\Gamma v$. Analogous trick of adding the artificial projector term in order to treat the neutral modes was devised already in [**DFKM96**]. In the present paper, we will apply this idea in more complicated situation of slowly evolving multi-pulses where the associated neutral modes also evolve slowly in time (and this evolution depends on the concrete choice of the multi-pulse trajectory, see Sections 7 and 8 below).

We now consider the analog of equation (3.18) on the positive semi-axis $[0, +\infty)$

$$\partial_t v + \mathcal{L}_\Gamma v + \mathbb{P}_\Gamma v = 0, \quad t \geq 0, \quad v\big|_{t=T} = v_0. \tag{3.24}$$

For simplicity, we consider only the spectrally stable case (where (2.25) holds).

COROLLARY 3.7. *Let the above assumptions hold and let, in addition, the initial pulse V be spectrally stable. Then, there exists a positive constant $\beta > 0$ such that, for every initial data $v_0 \in W^{2l(1-1/p),p}(\mathbb{R}^n)$, problem (3.24) possesses a unique solution and the following estimate holds:*

$$\|v\|_{W^{(1,2l),p}([T,T+1]\times\mathbb{R}^n)} \leq C\,e^{-\beta T}\,\|v_0\|_{W^{2l(1-1/p),p}(\mathbb{R}^n)} \tag{3.25}$$

where the constant C is independent of the concrete choice of Γ and v_0.

Indeed, due to (2.25), instead of (3.22) we have

$$\sigma(\mathcal{L}_{\Gamma_e} + \mathbb{P}_{\Gamma_e}) \cap \{\lambda \in \mathbb{C},\ \operatorname{Re}\lambda \in [-\delta, +\infty]\} = \varnothing \tag{3.26}$$

and, consequently, the exponentially unstable component of the exponential dichotomy for equation (3.24) vanishes and the whole phase space belongs to the exponentially stable component which implies (3.25) (here we implicitly used that the space $W^{2l(1-1/p),p}(\mathbb{R}^n)$ is a trace space for $W^{(1,2l),p}([0,1] \times \mathbb{R}^n)$ at $t = 0$, see [**LSU67**]).

In order to obtain the weighted analog of Proposition 3.5 and Corollary 3.7, we need the following Lemma which describes the behavior of the projectors $\mathbb{P}_\Gamma$ in weighted spaces.

LEMMA 3.8. *Let $\theta(x)$ be a weight function of a sufficiently small exponential growth rate $\varepsilon \le \varepsilon_0$. Then,*

$$\|\mathbb{P}_\Gamma v\|_{L^p_\theta(\mathbb{R}^n)} \le C\|v\|_{L^p_\theta(\mathbb{R}^n)} \tag{3.27}$$

where the constant C depends on C_θ, but is independent of the concrete form of the weight function θ and $\Gamma \in G_0$. Moreover, let $\theta_{\varepsilon,x_0}(x)$ be the special weights introduced in (3.3) *and let $M_{\theta_{\varepsilon,x_0}}$ be the multiplication operator on function θ_{ε,z_0}. Then,*

$$\|M_{\theta_{\varepsilon,x_0}} \circ \mathbb{P}_\Gamma \circ M_{\theta_{-\varepsilon,x_0}} - \mathbb{P}_\Gamma\|_{\mathcal{L}(L^p(\Omega),L^p(\Omega))} \le C\varepsilon \tag{3.28}$$

where the constant C is independent of ε and $x_0 \in \mathbb{R}^n$.

PROOF. We first note that, due to estimates (3.6) and (3.7), it is sufficient to verify estimates (3.27) only for the special weights (3.3). Moreover, since $L^p_\theta(\mathbb{R}^n) = M_\theta L^p(\mathbb{R}^n)$, in order to prove (3.27), it is sufficient to estimate the norm of $\mathbb{P}_{\Gamma,\varepsilon,x_0} := M_{\theta_{\varepsilon,x_0}} \circ \mathbb{P}_\Gamma \circ M_{\theta_{-\varepsilon,x_0}}$ in the non-weighted space. In order to do so, we represent $\mathbb{P}_{\Gamma,\varepsilon,x_0}$ and $K_{\Gamma,\varepsilon,x_0} := \mathbb{P}_{\Gamma,\varepsilon,x_0} - \mathbb{P}_\Gamma$ in the form of integral operators:

$$(\mathbb{P}_{\Gamma,\varepsilon,x_0}v)(y) = \int_{x\in\mathbb{R}^n} K(x,y)v(x)\,\mathrm{d}x,\ (K_{\Gamma,\varepsilon,x_0}v)(y) = \int_{x\in\mathbb{R}^n} K_1(x,y)v(x)\,\mathrm{d}x$$

where the (matrix) kernels K and K_1 have the following form

$$\begin{aligned} K(x,y) &:= [\sum_{i=1}^k \psi^i_\Gamma(x)\otimes\phi^i_\Gamma(y)]\theta_{-\varepsilon,x_0}(x)\theta_{\varepsilon,x_0}(y),\\ K_1(x,y) &:= [\sum_{i=1}^k \psi^i_\Gamma(x)\otimes\phi^i_\Gamma(y)](\theta_{-\varepsilon,x_0}(x)\theta_{\varepsilon,x_0}(y)-1). \end{aligned} \tag{3.29}$$

Without loss of generality, we may assume that $\Gamma := (0,\gamma)$ (the general case is reduced to that one by the appropriate space shift). Then, using the obvious inequality

$$\sqrt{|x_0|^2+1} - |z| \le \sqrt{|z-x_0|+1} \le |x_0| + \sqrt{|z|^2+1},$$

we obtain that

$$\begin{aligned} \theta_{-\varepsilon,x_0}(x)\theta_{\varepsilon,x_0}(y) &\le \mathrm{e}^{|\varepsilon|(|x|+|y|+1)}\\ |\theta_{-\varepsilon,x_0}(x)\theta_{\varepsilon,x_0}(y)-1| &\le |\varepsilon|(|x|+|y|+1)\,\mathrm{e}^{|\varepsilon|(|x|+|y|+1)} \end{aligned} \tag{3.30}$$

and, consequently, due to (2.32) (with $\xi = 0$), for sufficiently small $|\varepsilon|$, we have

$$|K(x,y)| \le \mathrm{e}^{-\alpha(|x|+|y|)/2},\quad |K_1(x,y)| \le C|\varepsilon|\,\mathrm{e}^{-\alpha(|x|+|y|)/2} \tag{3.31}$$

where the constant C is independent of ε and x_0. Using now that the $L^p - L^p$ norm of the integral operator is bounded by the following $L^p - L^q$ norm of its kernel K

$$\left(\int_{y\in\mathbb{R}^n}\left(\int_{x\in\mathbb{R}^n}|K(x,y)|^p\,\mathrm{d}x\right)^{q/p}\mathrm{d}y\right)^{1/q},\quad \frac{1}{p}+\frac{1}{q}=1 \tag{3.32}$$

together with estimates (3.31), we deduce the desired estimates (3.27) and (3.28) and finish the proof of Proposition 3.10. □

The following slight generalization of Lemma 3.8 (which is based on the simple observation that $(\mathbb{P}_\Gamma v)(x)$ decays exponentially like $\mathrm{e}^{-\alpha|x-\xi|}$ as $|x|\to\infty$ (due to estimates (2.32))), will be very important in the sequel in order to sum an infinite number of projectors $\mathbb{P}_{\Gamma_j}$ with different Γ_j.

COROLLARY 3.9. *Let the assumptions of Lemma 3.8 holds. Then, there exists a sufficiently small $\varepsilon_0 > 0$ such that, for every $\varepsilon < \varepsilon_0$ and every $x_0, y_0 \in \mathbb{R}^n$, we have the following estimates*

$$(3.33)\qquad \begin{aligned} \|\mathbb{P}_\Gamma v\|_{L^p_{e^{+\alpha/2|x-\xi|-\varepsilon|x-y_0|}}(\mathbb{R}^n)} &\leq C\|v\|_{L^p_{e^{-\varepsilon|x-y_0|}}(\mathbb{R}^n)} \\ \|\theta_{\varepsilon,x_0}\mathbb{P}_\Gamma(v\theta_{-\varepsilon,x_0}) - \mathbb{P}_\Gamma v\|_{L^p_{e^{\alpha/2|x-\xi|-\varepsilon|x-y_0|}}(\mathbb{R}^n)} &\leq C|\varepsilon| \cdot \|v\|_{L^p_{e^{-\varepsilon|x-y_0|}}(\mathbb{R}^n)} \end{aligned}$$

where the constant C is independent of p, x_0 and y_0.

Indeed, the proof of that estimates is completely analogous to the proof of Lemma 3.8 and we leave it to the reader.

We are now ready to formulate and prove the weighted analog of Proposition 3.5.

PROPOSITION 3.10. *Let the assumptions of Proposition 3.5 hold. Then, there exists a positive constant ε_0 depending only on p such that, for every weight function θ of exponential growth rate $\varepsilon \leq \varepsilon_0$ and every $h \in L^p_\theta(\mathbb{R}^{n+1})$, equation (3.18) possesses a unique solution and the following estimate holds:*

$$(3.34)\qquad \|v\|_{W^{(1,2l),p}_\theta(\mathbb{R}^{n+1})} \leq C_p\|h\|_{L^p_\theta(\mathbb{R}^{n+1})}$$

where the constant C_p depends on p and C_θ, but is independent on Γ, h and on the concrete form of the weight θ. Moreover, the analogous result holds for the spaces $L^p_{b,\theta}$ and for the adjoint equation (3.20).

PROOF. As in Proposition 3.3, it is sufficient to verify estimate (3.34) only for the special weights (3.3). It is however more convenient to use here the equivalent weights $\bar\theta_{\varepsilon,z_0}(t,x) := \theta_{\varepsilon,t_0}(t)\theta_{\varepsilon,x_0}(x)$, for $z_0 := (t_0, x_0) \in \mathbb{R}^{n+1}$. Indeed, let $\tilde v(t,x) = \bar\theta_{\varepsilon,z_0}(t,x)v(t,x)$. Then, analogously to Proposition 3.5, this function satisfies the following perturbed version of (3.18):

$$(3.35)\qquad \partial_t\tilde v + A_0\tilde v + F_\Gamma\tilde v + \mathbb{P}_\Gamma\tilde v = \mathbb{P}_\Gamma\tilde v - \theta_{\varepsilon,x_0}\mathbb{P}_\Gamma(\theta_{-\varepsilon,x_0}\tilde v) + \varepsilon\bar A_\varepsilon\tilde v + h\bar\theta_{\varepsilon,z_0}$$

where the operator $\bar A_\varepsilon$ satisfies (3.13) uniformly with respect to $z_0 \in \mathbb{R}^{n+1}$ and $\Gamma \in G_0$. Applying now estimate (3.19) to equation (3.35) and using (3.13) and (3.28), we obtain that, for sufficiently small $\varepsilon > 0$,

$$(3.36)\qquad \|\tilde v\|_{W^{(1,2l),p}(\mathbb{R}^{n+1})} \leq C_p\|h\bar\theta_{\varepsilon,z_0}\|_{L^p(\mathbb{R}^{n+1})}$$

where C_p is independent of Γ and $z_0 \in \mathbb{R}^{n+1}$. Estimate (3.34) for arbitrary weight can be obtained form (3.36) exactly as in Proposition 3.5. Proposition 3.10 is proven. □

COROLLARY 3.11. *Let the above assumptions hold and let, in addition, the initial pulse be spectrally stable. Then, there exists a positive constant β such that, for every weight $\theta \in C(\mathbb{R}^n)$ of sufficiently small exponential growth rate and every $v_0 \in W^{2l(1-1/p),p}_\theta(\mathbb{R}^n)$, equation (3.24) possesses a unique solution v and this solution satisfies:*

$$(3.37)\qquad \|v\|_{W^{(1,2l),p}_\theta([T,T+1]\times\mathbb{R}^n)} \leq C e^{-\beta T}\|v_0\|_{W^{2l(1-1/p),p}_\theta(\mathbb{R}^n)},\quad T \geq 0$$

where the constant C depends on C_θ, but is independent of the concrete choice of the weight θ and the initial data v_0. Moreover, the analogous result holds also for the spaces $W^{l,p}_{b,\theta}$.

Indeed, the proof of this estimate is completely analogous to the proof of Proposition 3.10, only we need to use additionally Corollary 3.7 in order to solve problem (3.18) on a semi-axis with inhomogeneous initial data.

We conclude this section by preparing some technical tools which allow us to deal with the infinite sums of operators $F_{\Gamma_j} + \mathbb{P}_{\Gamma_j}$ for "well-separated" sequences Γ_j.

LEMMA 3.12. *For each $L > 0$ and $\varepsilon > 0$ there exists a positive constant $C = C(L, \varepsilon)$ such that, for every sequence $\Xi = \{\xi_j\}_{j\in\mathbb{N}}$ of vectors $\xi_j \in \mathbb{R}^n$ satisfying*

$$|\xi_i - \xi_j| > 2L, \quad i, j \in \mathbb{N}, \quad i \neq j, \tag{3.38}$$

the function

$$R_{\varepsilon,\Xi} : \mathbb{R}^n \to \mathbb{R}, \quad R_{\varepsilon,\Xi}(x) := \sum_{i=1}^{\infty} \mathrm{e}^{-\varepsilon|x-\xi_j|} \tag{3.39}$$

is well-defined and possesses the following estimate:

$$R_{\varepsilon,\Xi}(x) \le C(1 + [\mathrm{dist}(x, \Xi)]^n)\, \mathrm{e}^{-\varepsilon\, \mathrm{dist}(x,\Xi)}\,. \tag{3.40}$$

PROOF. Indeed, since all of ξ_i satisfy (3.38), then the number $N_x(s)$ of points ξ_i belonging to the ball $B_x(s)$, $s \in \mathbb{N}$, can be estimated via

$$N_x(s) \le C(\frac{s}{L} + 1)^n \tag{3.41}$$

where the constant C is independent on s, L, Ξ and x. Therefore,

$$R_{\varepsilon,\Xi}(x) \le \sum_{s=1}^{\infty} N_x(s)\, \mathrm{e}^{-\varepsilon(s-1)} \le C \sum_{s=1}^{\infty} (s+1)^n\, \mathrm{e}^{-\varepsilon(s-1)} \le C_\varepsilon. \tag{3.42}$$

Thus, the function (3.39) is indeed well-defined. In order to prove (3.40), we set $M := \mathrm{dist}(x, \Xi)$. Then, obviously

$$N_x(s) = 0, \quad s = 1, \cdots, [M] - 1 \tag{3.43}$$

and, consequently, (3.43) reads

$$\begin{aligned} R_{\varepsilon,\Xi}(x) \le \sum_{s=[M]}^{\infty} N_x(s)\, \mathrm{e}^{-\varepsilon(s-1)} \le C \sum_{s=[M]}^{\infty} (s+1)^n\, \mathrm{e}^{-\varepsilon(s-1)} \le \\ \le C_1\, \mathrm{e}^{-\varepsilon M} \sum_{s=1}^{\infty} (s^n + M^n)\, \mathrm{e}^{-\varepsilon s} \le C_2(1 + M^n)\, \mathrm{e}^{-\varepsilon M}\,. \end{aligned} \tag{3.44}$$

Lemma 3.12 is proven. □

The following proposition is the main technical tool for dealing with infinite sums of well-separated operators.

PROPOSITION 3.13. *Let Ξ be a sequence satisfying* (3.38) *and let H_i, $i = \overline{1, \infty}$ be a sequence of operators (not necessarily linear) which satisfy*

$$\|H_i v\|_{L^p_{\mathrm{e}^{+\beta|x-\xi_i|-\varepsilon_0|x-x_0|}}(\mathbb{R}^n)} \le K \|v\|_{L^p_{\mathrm{e}^{-\varepsilon_0|x-x_0|}}(\mathbb{R}^n)} \tag{3.45}$$

for some positive β and ε_0 and a constant K independent of $i \in \mathbb{N}$ and $x_0 \in \mathbb{R}^n$. Then, for every $\varepsilon \le \varepsilon_0/2$, the operator $H := \sum_{i=1}^{\infty} H_i$ acts from $L^p_\theta(\mathbb{R}^n)$ to $L^p_\theta(\mathbb{R}^n)$ and the following estimate holds:

$$\|Hv\|_{L^p_\theta(\mathbb{R}^n)} \le C_{\varepsilon_0,\beta} K \|h\|_{L^p_\theta(\mathbb{R}^n)} \tag{3.46}$$

where the constant C depends on ε_0, β and C_θ, but is independent of Ξ, H_i, h, K and $L \geq L_0$. Moreover, the analog of (2.46) *for the spaces $L^p_{b,\theta}$ holds as well.*

PROOF. Indeed, estimate (3.45) together with (3.5) imply that

$$\|H_i v\|_{L^p(B^1_{x_0})} \leq CK\,\mathrm{e}^{-\beta|x_0-\xi_i|}\left(\int_{x\in\mathbb{R}^n}\mathrm{e}^{-p\varepsilon|x-x_0|}\,\|v\|^p_{L^p(B^1_x)}\,dx\right)^{1/p} \tag{3.47}$$

where the constant C is independent of K, Ξ and i. Using estimate (3.47) together with (3.40), we infer

$$\|Hv\|_{L^p(B^1_{x_0})} \leq \sum_{i=1}^{\infty}\|H_i v\|_{L^p(B^1_{x_0})} \leq \tag{3.48}$$

$$\leq CK\cdot R_{\beta,\Xi}(x_0)\left(\int_{x\in\mathbb{R}^n}\mathrm{e}^{-p\varepsilon|x-x_0|}\,\|v\|^p_{L^p(B^1_x)}\,dx\right)^{1/p} \leq$$

$$\leq C_\beta K\left(\int_{x\in\mathbb{R}^n}\mathrm{e}^{-p\varepsilon|x-x_0|}\,\|v\|^p_{L^p(B^1_x)}\,dx\right)^{1/p}.$$

Multiplying this inequality by $\theta(x_0)$, taking supremum over $x_0 \in \mathbb{R}^n$ and using (3.7), we obtain the analog of (3.46) for the spaces $L^p_{b,\theta}(\mathbb{R}^n)$. In order to obtain (3.46) for the case of space L^p_θ ($p < \infty$!), we rewrite (3.48) in the equivalent form:

$$\|Hv\|^p_{L^p(B^1_{x_0})} \leq (C_\beta K)^p\int_{x\in\mathbb{R}^n}\mathrm{e}^{-\varepsilon p|x-x_0|}\,\|v\|^p_{L^p(B^1_x)}\,dx. \tag{3.49}$$

Multiplying this relation by $\theta^p(x_0)$ integrating over $x_0 \in \mathbb{R}^n$ and using (3.6), we deduce (3.46) and finish the proof of Proposition 3.13. □

COROLLARY 3.14. *Let the operators A_0 and Φ satisfy all of the assumptions of Section 2 and let, in addition, $\vec{\Gamma} := \{\Gamma_j = (\xi_j, \gamma_j)\}_{j=1}^{\infty} \in G_0^{\mathbb{N}}$ be a sequence of transformations such that the associated sequence $\Xi := \{\xi_j\}_{j=1}^{\infty}$ satisfies* (3.38). *Then the operator*

$$\mathbb{F}_{\vec{\Gamma}} := \sum_{j=1}^{\infty}(F_{\Gamma_j} + \mathbb{P}_{\Gamma_j}) \tag{3.50}$$

is well-defined on $W^{2l-1,p}(\mathbb{R}^n)$ and, there exists $\varepsilon_0 > 0$ (which is independent of Γ_j) such that, for every weight function θ of exponential growth rate $\varepsilon \leq \varepsilon_0$, we have

$$\|\mathbb{F}_{\vec{\Gamma}} v\|_{W^{2l-1,p}_\theta(\mathbb{R}^n)} \leq C\|v\|_{L^p_\theta(\mathbb{R}^n)} \tag{3.51}$$

where the constant C depends on A_0 and Φ and C_θ, but is independent of $L \geq L_0$ and the sequence Γ_j. Analogously, the operator

$$\mathbb{K} := \sum_{j=1}^{\infty}[M_{\theta_{\varepsilon,x_0}} \circ \mathbb{P}_{\Gamma_j} \circ M_{\theta_{-\varepsilon,x_0}} - \mathbb{P}_{\Gamma_j}] \tag{3.52}$$

is well defined on $L^p(\mathbb{R}^n)$ for all $x_0 \in \mathbb{R}^n$ and $\varepsilon \leq \varepsilon_0$ and

$$\|\mathbb{K}v\|_{L^p_\theta(\mathbb{R}^n)} \leq C|\varepsilon|\|v\|_{L^p_\theta(\mathbb{R}^n)} \tag{3.53}$$

where the constant C is independent of ε, Γ_i and x_0. Moreover, the analogous estimates hold also for the spaces $L^p_{b,\theta}$.

PROOF. Indeed, since all of the coefficients of the differential operator F_{Γ_j} satisfy estimate (2.32), then we have

$$\|F_{\Gamma_j} v\|_{L^p_{e^{+\beta|x-\xi_j|-\varepsilon|x-x_0|}}(\mathbb{R}^n)} \leq C\|v\|_{W^{2l-1,p}_{e^{-\varepsilon|x-x_0|}}(\mathbb{R}^n)} \tag{3.54}$$

where the constant C is independent of Γ_j and $x_0 \in \mathbb{R}^n$. All of the assertions of Corollary 3.14 follow now from Proposition 3.13, estimate (3.54) and Corollary 3.9. Corollary 3.14 is proven. □

REMARK 3.15. As it follows from the proof of Proposition 3.13, the result remains valid if we replace the weights $e^{-\beta|x-\xi_i|}$ by arbitrary weights $\theta_i(x)$ satisfying

$$\sum_{i=1}^{\infty} \theta_i(x) \leq C < \infty \tag{3.55}$$

where the constant C is independent of x. We will use below this simple observation in the situation, when we have space-time integration in (3.45) and (3.46) (over $(t,x) \in \mathbb{R}^{n+1}$), but the weights $\theta_i(x) = e^{\beta|x-\xi_i|}$ will depend only on the spatial variable.

4. The multi-pulse manifold: general structure

In this section, we start to study the manifold of non-interacting (standing) pulses. To this end, we first need to define and study the space of sequences $\vec{\Gamma} = \{\Gamma_j\}_{j=1}^{\infty}$ of admissible pulse configurations.

DEFINITION 4.1. Let L be a sufficiently large positive number. Let us define the space $\tilde{\mathbb{B}}(L)$ of $2L$-separated sequences as follows

$$\tilde{\mathbb{B}}(L) := \{\vec{\Gamma} := \{(\xi_j, \gamma_j)\}_{j=1}^{\infty} \in (G_0)^{\mathbb{N}}, \ \operatorname{sep}(\vec{\Gamma}) > 2L\}. \tag{4.1}$$

where

$$\operatorname{sep}(\vec{\Gamma}) := \inf\{|\xi_i - \xi_j|, \ \ i \neq j\}.$$

Moreover, we set $\Sigma = \{\xi_j\}_{j\in\mathbb{N}}$ and define the associated semi-metric on $\tilde{\mathbb{B}}(L)$ as follows:

$$d(\vec{\Gamma}^1, \vec{\Gamma}^2) := \sup_{j\in\mathbb{N}} \inf_{i\neq j} d(\vec{\Gamma}^1_j, \vec{\Gamma}^2_i) \tag{4.2}$$

and

$$d^s(\vec{\Gamma}^1, \vec{\Gamma}^2) := \max\{d(\vec{\Gamma}^1, \vec{\Gamma}^2), d(\vec{\Gamma}^2, \vec{\Gamma}^1)\} \tag{4.3}$$

where $d(\cdot,\cdot)$ is some bounded metric on the G_0 (e.g., $d(\Gamma^1, \Gamma^2) = \min\{1, \|\Gamma^1 - \Gamma^2\|\}$ where $\|\cdot\|$ is the norm on the group G_0 introduced in Section 2.).

We note that $d^s(\vec{\Gamma}^1, \vec{\Gamma}^2) = 0$ if and only if there exists an infinite permutation $\pi : \mathbb{N} \to \mathbb{N}$ such that $\vec{\Gamma}^1 = \pi\vec{\Gamma}^2 = \{\Gamma^2_{\pi(j)}\}_{j=1}^{\infty}$. Moreover, obviously, $d^s(\pi_1\vec{\Gamma}^1, \pi_2\vec{\Gamma}^2) = d^{sym}(\vec{\Gamma}^1, \vec{\Gamma}^2)$ for every two permutations $\pi_1, \pi_2 \in S_\infty$ and, consequently, (4.3) gives a metric on a factor space

$$\mathbb{B}(L) := \tilde{\mathbb{B}}(L)/S_\infty. \tag{4.4}$$

It is not difficult to verify that $\mathbb{B}(L)$ is a smooth manifold over the B-space $l^\infty = l^\infty(\mathbb{R}^k)$ with the standard norm $\|r\|_{l^\infty} := \sup_{j\in\mathbb{N}} \|r_j\|_{\mathbb{R}^k}$. Indeed, for every $\vec{\Gamma}^0 \in \mathbb{B}(L)$, the set

$$\mathcal{B}_{r_0}(\vec{\Gamma}^0) := \{\vec{\Gamma} \in \tilde{\mathbb{B}}(L),\ \ \|\vec{\Gamma}^0 - \vec{\Gamma}\|_\infty := \sup_{j\in\mathbb{N}} \|\Gamma_j^0 - \Gamma_j\| < r_0\} \tag{4.5}$$

is an open neighborhood of $\vec{\Gamma}^0$ in $\mathbb{B}(L)$ if $r_0 > 0$ is small enough. Moreover, analogously to Proposition 2.2, the map $\alpha \to \vec{\Gamma}(\alpha)$ from l^∞ to $[G_0]^{\mathbb{N}}$ defined by

$$\vec{\Gamma}(\alpha) := \{\Gamma_j(\alpha_j)\}_{j\in\mathbb{N}},\ \ \Gamma(\alpha_j) := \Gamma_j^0 \circ \exp(\alpha_j \cdot v),\ \ \alpha = \{\alpha_j\}_{j\in\mathbb{N}} \in l^\infty(\mathbb{R}^k), \tag{4.6}$$

where $v = \{v^i\}_{i=1}^k$ is a fixed basis in the Lie algebra $\mathcal{G}$, is a local diffeomorphism and, consequently, gives the natural local coordinates on $\mathbb{B}(L)$ near $\vec{\Gamma}^0$. Thus, the whole neighborhood $\mathcal{B}_{r_0}(\vec{\Gamma}^0)$ belongs to the same coordinate neighborhood if $r_0 > 0$ is small enough and, due to the uniformity proven in Proposition 2.2, the radius r_0 can be chosen uniformly with respect to $\vec{\Gamma}^0 \in \mathbb{B}(L+\delta)$, $\delta > 0$ and the norms of coordinate diffeomorphisms will be also independent of $\vec{\Gamma}^0$.

In the sequel we will identify (everywhere where it does not lead to misunderstandings) the element $\vec{\Gamma} \in \mathbb{B}(L)$ with its proper coordinates and write $\Gamma_j^i \in l^\infty$ instead of $\alpha_j^i \in l^\infty$.

We note that, obviously, $\mathbb{B}(L) \subset \mathbb{B}(L-\delta)$ if $\delta > 0$. In the sequel, we will also need the "boundary" $\partial\mathbb{B}(L)$ which is defined via

$$\partial\mathbb{B}(L) := \{\vec{\Gamma} \in \mathbb{B}(L-\delta),\ \delta > 0,\ \ \mathrm{sep}(\vec{\Gamma}) = 2L\}. \tag{4.7}$$

Moreover, we also need to introduce the local topology on the space $\mathbb{B}(L)$.

DEFINITION 4.2. We say that a sequence $\vec{\Gamma}^n \in \mathbb{B}(L)$ converges locally to some $\vec{\Gamma} \in \mathbb{B}(L)$ if, for every $R > 0$

$$\lim_{n\to\infty} \mathrm{dist}_{\mathrm{sH}}(\{\vec{\Gamma}^n\}_R, \{\vec{\Gamma}\}_R) = 0 \tag{4.8}$$

where $\{\vec{\Gamma}\}_R \subset \mathbb{R}^{N\times N}$ is a set of all components $\Gamma_j = (\xi_j, \gamma_j)$ of $\vec{\Gamma}$ satisfying $|\xi_j| < R$ and $\mathrm{dist}_{\mathrm{sH}}$ is a symmetric Hausdorff distance between sets in $\mathbb{R}^{N\times N}$.

For every $\gamma > 0$, we introduce also the following "quasimetric" on $\mathbb{B}(L)$:

$$d_\gamma^s(\vec{\Gamma}^1, \vec{\Gamma}^2) := \max\{d_\gamma(\vec{\Gamma}^1, \vec{\Gamma}^2), d_\gamma(\vec{\Gamma}^2, \vec{\Gamma}^1)\}$$

where

$$d_\gamma(\vec{\Gamma}^1, \vec{\Gamma}^2) := \sup_{j\in\mathbb{N}}\{e^{-\gamma|\xi_j^1|} \inf_{i\in\mathbb{N}} d(\vec{\Gamma}_j^1, \vec{\Gamma}_i^2)\} \tag{4.9}$$

(compare with (4.2) and (4.3)). Then, as not difficult to see, the local convergence is equivalent to the convergence with respect to d_γ^s.

We define also the pulse map $\mathbb{V} : \mathbb{B}(L) \to [L^\infty(\mathbb{R}^n)]^m$ by the following expression:

$$\mathbb{V}(\vec{\Gamma}) = \mathbb{V}_{\vec{\Gamma}} := \sum_{j=1}^{\infty} V_{\Gamma_j} \tag{4.10}$$

where $V_{\Gamma_j} = \mathcal{T}_{\Gamma_j} V$ is a Γ_j-shift of the initial pulse V.

We are now ready to define the central object of our paper, the multi-pulse manifold $\mathbb{P}(L)$ as follows:

$$\mathbb{P}(L) := \mathbb{V}(\mathbb{B}(L)) \tag{4.11}$$

which is naturally endowed by the topology, induced by the embedding $\mathbb{P}(L) \subset [L^\infty(\mathbb{R}^n)]^m$.

The next proposition shows that this set is indeed a smooth submanifold diffeomorphic to $\mathbb{B}(L)$ (and clarifies our choice of the topologies on $\mathbb{B}(L)$).

PROPOSITION 4.3. *Let the pulses V_{Γ_j} satisfy the assumptions of Section 2. Then, for a sufficiently large $L > 0$, the set $\mathbb{P}(L)$ is a smooth submanifold of $[L^\infty(\mathbb{R}^n)]^m$ and the function (4.10) is a diffeomorphism between $\mathbb{P}(L)$ and $\mathbb{B}(L)$. Moreover, there exists positive constants C_0 and γ_0 independent of $L \geq L_0$ such that, for every $\gamma \in [0, \gamma_0]$*

$$C_0^{-1} d^s_\gamma(\vec{\Gamma}^1, \vec{\Gamma}^2) \leq \|\mathbb{V}_{\vec{\Gamma}^1} - \mathbb{V}_{\vec{\Gamma}^2}\|_{L^\infty_{e^{-\gamma|x|}}(\mathbb{R}^n)} \leq C_0 d^s_\gamma(\vec{\Gamma}^1, \vec{\Gamma}^2) \tag{4.12}$$

for all $\vec{\Gamma}^1, \vec{\Gamma}^2 \in \mathbb{B}(L)$.

PROOF. Let us introduce some notations which are useful for weighted estimates of multi-pulses. Let $\vec{\Gamma}^1, \vec{\Gamma}^2 \in \mathbb{B}(L)$ and let $0 < r < L/2$. We introduce the following subsets of $\mathbb{N}$:

$$\mathbb{N}^1_c(r) := \{j \in \mathbb{N},\ \ \exists i \in \mathbb{N},\ \|\Gamma^1_j - \Gamma^2_i\| \leq r\},\ \ \mathbb{N}^1_o(r) := \mathbb{N}\backslash\mathbb{N}^1_o(r) \tag{4.13}$$

and the analogous subsets $\mathbb{N}^2_c(r)$ and $\mathbb{N}^2_o(r)$. Then, up to the renumeration, we may assume that

$$\mathbb{N}^1_c(r) = \mathbb{N}^2_c(r) := \mathbb{N}_c(r) \quad \text{and} \quad \|\Gamma^1_j - \Gamma^2_j\| \leq r,\ \ j \in \mathbb{N}_c(r). \tag{4.14}$$

Then, the equivalent weighted distance D^s_γ between $\vec{\Gamma}^i$ can be now computed via

$$D^s_\gamma(\vec{\Gamma}^1, \vec{\Gamma}^2) = \max\{\sup_{j\in\mathbb{N}^1_o(r)}\{e^{-\gamma|\xi^1_j|}\},\ \sup_{j\in\mathbb{N}^2_o(r)}\{e^{-\gamma|\xi^2_j|}\},\ \sup_{j\in\mathbb{N}_c(r)}\{e^{-\gamma|\xi^1_j|}\,\|\Gamma^1_j - \Gamma^2_j\|\}\}. \tag{4.15}$$

Indeed, it is not difficult to verify that

$$C^{-1} d^s_\gamma(\vec{\Gamma}^1, \vec{\Gamma}^2) \leq D^s_\gamma(\vec{\Gamma}^1, \vec{\Gamma}^2) \leq C d^s_\gamma(\vec{\Gamma}^1, \vec{\Gamma}^2)$$

where the constant C depends on r, but is independent of L. We also set $\vec{\Gamma}^i_o := \{\Gamma^i_j\}_{j\in\mathbb{N}^i_o(r)}$ and $\vec{\Gamma}^i_c := \{\vec{\Gamma}^i_j\}_{j\in\mathbb{N}_c(r)}$, $i = 1, 2$. Then, obviously

$$\mathbb{V}_{\vec{\Gamma}^i} = \mathbb{V}_{\vec{\Gamma}^i_o} + \mathbb{V}_{\vec{\Gamma}^i_c},\quad i = 1, 2. \tag{4.16}$$

Let us first prove the inequality in the right-hand side of (4.12). Indeed, let $r = 1$. Then, due to (4.16)

$$\|\mathbb{V}_{\vec{\Gamma}^1} - \mathbb{V}_{\vec{\Gamma}^2}\|_{L^\infty_{e^{-\gamma|x|}}(\mathbb{R}^n)} \leq \|\mathbb{V}_{\vec{\Gamma}^1_o}\|_{L^\infty_{e^{-\gamma|x|}}(\mathbb{R}^n)} + \\ + \|\mathbb{V}_{\vec{\Gamma}^2_o}\|_{L^\infty_{e^{-\gamma|x|}}(\mathbb{R}^n)} + \|\mathbb{V}_{\vec{\Gamma}^1_c} - \mathbb{V}_{\vec{\Gamma}^2_c}\|_{L^\infty_{e^{-\gamma|x|}}(\mathbb{R}^n)}. \tag{4.17}$$

Then, due to (2.10) and (2.11) and Lemma 3.12, for $\gamma \leq \alpha/2$, we have

$$\|\mathbb{V}_{\vec{\Gamma}^1_o}\|_{L^\infty_{e^{-\gamma|x|}}(\mathbb{R}^n)} \leq \sup_{x\in\mathbb{R}^n} \sum_{j\in\mathbb{N}^1_o} e^{-\alpha|x-\xi^1_j|-\gamma|x|} \leq \\ \leq \sup_{j\in\mathbb{N}^1_o}\{e^{-\gamma|\xi_j|}\} \sup_{x\in\mathbb{R}^n} R_{\alpha-\gamma,\Sigma^1}(x) \leq C \sup_{j\in\mathbb{N}^1_o}\{e^{-\gamma|\xi_j|}\}. \tag{4.18}$$

The second term in the right-hand side of (4.17) is completely analogous and for estimating the third one we need to use additionally that

$$|V_{\Gamma_j^1}(x) - V_{\Gamma_j^2}(x)| \le C\,e^{-\alpha|x-\xi_j^1|}\,\|\Gamma_j^1 - \Gamma_j^2\| \tag{4.19}$$

where the constant C is independent of j and of $\Gamma_j^1, \Gamma_j^2 \in G_0$ such that $\|\Gamma_j^1 - \Gamma_j^2\| \le 1$ (which is also and immediate corollary of (2.10) and (2.11)). Thus, the right-hand side of inequality (4.12) is verified.

Let us now verify the left-hand side of (4.12). To this end, we will use again splitting (4.13)–(4.14), but now with sufficiently large $r < L/2$ which will be fixed below. Let us first consider some $j_0 \in \mathbb{N}_o^1$. Then, by definition,

$$\operatorname{dist}(\xi_j^1, \Xi^2) \ge 2r' := r - C_1 \tag{4.20}$$

(here C_1 is the norm of the compact group $\bar{G}_0$ realized as a subset of $\mathbb{R}^{N\times N}$ in the metric $\|\cdot\|$ defined in Section 2) and

$$\begin{aligned}\|\mathbb{V}_{\vec\Gamma^1} - \mathbb{V}_{\vec\Gamma^2}\|_{L^\infty_{e^{-\gamma|x|}}(\mathbb{R}^n)} \ge \|\mathbb{V}_{\vec\Gamma^1} - \mathbb{V}_{\vec\Gamma^2}\|_{L^\infty_{e^{-\gamma|x|}}(B^{r'}_{\xi^1_{j_0}})} \ge \|V_{\Gamma^1_{j_0}}\|_{L^\infty_{e^{-\gamma|x|}}(B^{r'}_{\xi^1_{j_0}})} - \\ - \|\sum_{j\in\mathbb{N}_o^1, j\ne j_0}^{\infty} V_{\Gamma_j^1}\|_{L^\infty_{e^{-\gamma|x|}}(B^{r'}_{\xi^1_{j_0}})} - \|\mathbb{V}_{\vec\Gamma^1_c} - \mathbb{V}_{\vec\Gamma^2_c}\|_{L^\infty_{e^{-\gamma|x|}}(B^{r'}_{\xi^1_{j_0}})} - \|\mathbb{V}_{\vec\Gamma^2_o}\|_{L^\infty_{e^{-\gamma|x|}}(B^{r'}_{\xi^1_{j_0}})}.\end{aligned} \tag{4.21}$$

Furthermore, according to Lemma 3.12, and estimates (2.10) and (2.11), we have

$$\begin{aligned}\|\sum_{j\in\mathbb{N}_o^1, j\ne j_0}^{\infty} V_{\Gamma_j^1}\|_{L^\infty_{e^{-\gamma|x|}}(B^{r'}_{\xi^1_{j_0}})} \le \\ \sup_{|x-\xi^1_{j_0}|\le r'} \Big\{\sum_{j\in\mathbb{N}_o^1(r), j\ne j_0} e^{-(\alpha-\gamma)|x-\xi_j^1|}\Big\} \sup_{j\in\mathbb{N}_o^1(r)} \{e^{-\gamma|\xi_j^1|}\} \le C\,e^{-\alpha(L-r')/2}\, D^s_\gamma(\vec\Gamma^1, \vec\Gamma^2)\end{aligned} \tag{4.22}$$

where we have implicitly assumed that $\gamma \le \alpha/2$. Analogously,

$$\begin{aligned}\|\mathbb{V}_{\vec\Gamma^1_c} - \mathbb{V}_{\vec\Gamma^2_c}\|_{L^\infty_{e^{-\gamma|x|}}(B^{r'}_{\xi^1_{j_0}})} \le C\,e^{-\alpha(L-r')/2}\, D^s_\gamma(\vec\Gamma^1, \vec\Gamma^2), \\ \|\mathbb{V}_{\vec\Gamma^2_o}\|_{L^\infty_{e^{-\gamma|x|}}(B^{r'}_{\xi^1_{j_0}})} \le C\,e^{-\alpha r'/2}\, D^s_\gamma(\vec\Gamma^1, \vec\Gamma^2)\end{aligned} \tag{4.23}$$

(here we have implicitly used (4.20)). Inserting these estimates to (4.21), we arrive at

$$\|\mathbb{V}_{\vec\Gamma^1} - \mathbb{V}_{\vec\Gamma^2}\|_{L^\infty_{e^{-\gamma|x|}}(\mathbb{R}^n)} \ge \|V_{\Gamma^1_{j_0}}\|_{L^\infty_{e^{-\gamma|x|}}(B^{r'}_{\xi^1_{j_0}})} - C(e^{-\alpha(L-2r')/4} + e^{-\alpha r'/4}) D^s_\gamma(\vec\Gamma^1, \vec\Gamma^2).$$

Using now (2.43), we obtain

$$\|V_{\Gamma^1_{j_0}}\|_{L^\infty_{e^{-\gamma|x|}}(B^{r'}_{\xi^1_{j_0}})} \ge \kappa\, e^{-\gamma|\xi^1_{j_0}|}$$

for some positive κ (which is independent of $r' \ge R_0$). Assuming now that L is large enough and fixing $r' \ge R_0$ in such way that

$$C(e^{-\alpha(L-r')/2} + e^{-\alpha r'/2}) \le \kappa/2,$$

we will have

$$\|\mathbb{V}_{\vec\Gamma^1} - \mathbb{V}_{\vec\Gamma^2}\|_{L^\infty_{e^{-\gamma|x|}}(\mathbb{R}^n)} \ge \kappa\, e^{-\gamma|\xi^1_{j_0}|} - \kappa/2 D^s_\gamma(\vec\Gamma^1, \vec\Gamma^2)$$

and, taking the supremum over $j_0 \in \mathbb{N}_o^1$,

$$\|\mathbb{V}_{\vec{\Gamma}^1} - \mathbb{V}_{\vec{\Gamma}^2}\|_{L^\infty_{e^{-\gamma|x|}}(\mathbb{R}^n)} \geq \kappa \sup_{j\in\mathbb{N}_o^1}\{e^{-\gamma|\xi_j^1|}\} - \kappa/2D^s_\gamma(\vec{\Gamma}^1,\vec{\Gamma}^2). \tag{4.24}$$

Arguing analogously, we establish also that

$$\begin{aligned} &\|\mathbb{V}_{\vec{\Gamma}^1} - \mathbb{V}_{\vec{\Gamma}^2}\|_{L^\infty_{e^{-\gamma|x|}}(\mathbb{R}^n)} \geq \kappa \sup_{j\in\mathbb{N}_c}\{e^{-\gamma|\xi_j|}\,\|\Gamma_j^1 - \Gamma_j^2\|\} - \kappa/2D^s_\gamma(\vec{\Gamma}^1,\vec{\Gamma}^2),\\ &\|\mathbb{V}_{\vec{\Gamma}^1} - \mathbb{V}_{\vec{\Gamma}^2}\|_{L^\infty_{e^{-\gamma|x|}}(\mathbb{R}^n)} \geq \kappa \sup_{j\in\mathbb{N}_o^2}\{e^{-\gamma|\xi_j^2|}\} - \kappa/2D^s_\gamma(\vec{\Gamma}^1,\vec{\Gamma}^2). \end{aligned} \tag{4.25}$$

Combining (4.24) and (4.25), we deduce

$$\|\mathbb{V}_{\vec{\Gamma}^1} - \mathbb{V}_{\vec{\Gamma}^2}\|_{L^\infty_{e^{-\gamma|x|}}(\mathbb{R}^n)} \geq \kappa/2D^s_\gamma(\vec{\Gamma}^1,\vec{\Gamma}^2)$$

which finishes the proof of estimate (4.12).

In particular, taking $\gamma = 0$ in that estimate, we see that $\mathbb{V}$ is a Lipschitz continuous isomorphism between $\mathbb{B}(L)$ and $\mathbb{P}(L)$ and, consequently, $\mathbb{P}(L)$ is a submanifold of $[L^\infty(\mathbb{R}^n)]^m$. The differentiability of $\mathbb{V}$ can be verified completely analogously and we left it to the reader. Thus, Proposition 4.3 is proven. □

The next corollary shows that all Sobolev's norms are equivalent on the manifold $\mathbb{P}(L)$.

COROLLARY 4.4. *Let the assumptions of Proposition 4.3 holds and let* $1 \leq p \leq \infty$, $l \in [0,\infty)$ *such that* $\frac{l}{n} - \frac{1}{p} > 0$. *Then, there exists* $C = C(p,l)$ *such that*

$$C^{-1}\|\mathbf{m}_1 - \mathbf{m}_2\|_{L^\infty_{e^{-\gamma|x|}}(\mathbb{R}^n)} \leq \|\mathbf{m}_1 - \mathbf{m}_2\|_{W^{l,p}_{b,e^{-\gamma|x|}}(\mathbb{R}^n)} \leq C\|\mathbf{m}_1 - \mathbf{m}_2\|_{L^\infty_{e^{-\gamma|x|}}(\mathbb{R}^n)} \tag{4.26}$$

for every $\mathbf{m}_1, \mathbf{m}_2 \in \mathbb{P}(L)$

PROOF. Indeed, since $W^{l,p} \subset L^\infty$, then taking in account Proposition 4.3, it is sufficient to verify only that

$$\|\mathbb{V}_{\vec{\Gamma}^1} - \mathbb{V}_{\vec{\Gamma}^2}\|_{W^{l,\infty}_{e^{-\gamma|x|}}(\mathbb{R}^n)} \leq C_l d^s_\gamma(\vec{\Gamma}^1,\vec{\Gamma}^2)$$

and this inequality can be verified exactly as in the case $l = 0$ (see Proposition 4.3). □

REMARK 4.5. The last proposition allows us to fix a convenient uniformly smooth atlas (in the sense that all transfer maps are uniformly smooth) of local coordinates on $\mathbb{P}(L)$. Indeed, let $\alpha \to \vec{\Gamma}(\alpha)$ be the local coordinates near $\vec{\Gamma}^0 \in \mathbb{B}(L)$ introduced in Definition 4.1. Then, due to Propositions 4.3 with $\gamma = 0$ and 2.2, we have that the whole neighborhood

$$\mathcal{O}_{r_0'}(\mathbb{V}_{\vec{\Gamma}^0}) := \{\mathbf{m} \in \mathbb{P}(L),\ \ \|\mathbf{m} - \mathbb{V}_{\vec{\Gamma}^0}\|_{L^\infty(\mathbb{R}^n)} \leq r_0'\}$$

(where $r_0' = r_0'(r_0) > 0$ is sufficiently small positive number, independent of $\vec{\Gamma}^0 \in \mathbb{B}(L+\delta)$ belongs to the same coordinate neighborhood of that atlas and, for sufficiently large L and every $\vec{\Gamma}^1 = \vec{\Gamma}(\alpha^1), \vec{\Gamma}^2 = \vec{\Gamma}(\alpha^2) \in \mathcal{O}_{r_0'}(\mathbb{V}_{\vec{\Gamma}^0})$,

$$C_0^{-1}\|\alpha^1 - \alpha^2\|_{l^\infty} \leq \|\mathbb{V}_{\vec{\Gamma}(\alpha^1)} - \mathbb{V}_{\vec{\Gamma}(\alpha^2)}\|_{L^\infty(\mathbb{R}^n)} \leq C_0\|\alpha^1 - \alpha^2\|_{l^\infty} \tag{4.27}$$

where the constant C_0 is independent of the concrete choice of $\vec{\Gamma}^0 \in \mathbb{B}(L)$. Thus, the map $\alpha \to \mathbb{V}_{\vec{\Gamma}(\alpha)}$ where $\vec{\Gamma}(\alpha)$ is defined by (4.6) gives indeed a natural local coordinates near $\mathbb{V}_{\vec{\Gamma}_0}$. In the sequel, the local coordinates on $\mathbb{P}(L)$ will always mean the local coordinates belonging to the above described atlas.

REMARK 4.6. We define also the boundary $\partial\mathbb{P}(L) := \mathbb{V}(\partial\mathbb{B}(L))$. Then, Proposition 4.3 shows also that the map $\mathbb{V}$ is a Lipschitz continuous isomorphism of $\partial\mathbb{P}(L)$ and $\partial\mathbb{B}(L)$.

Let us now consider the tangent space $T_{\mathbf{m}}\mathbb{P}(L)$ to the manifold $\mathbb{P}(L)$ at point $\mathbf{m} = \mathbb{V}_{\vec{\Gamma}}$ endowed by the metric of $[L^\infty(\mathbb{R}^n)]^m$. Obviously, every element $\mathrm{v} \in T_{\mathbf{m}}\mathbb{P}(L)$ has a form

$$\mathrm{v} = \mathrm{v}(a) = \sum_{j=1}^{\infty}\sum_{i=1}^{k} a_j^i \phi_{\Gamma_j}^i, \quad a = \{a_j^i\} \in l^\infty(\mathbb{R}^k),$$

see (2.37) and (4.10). Then, analogously to Proposition 4.3, one establishes that

$$C^{-1}\|a\|_{l^\infty} \le \|\,\mathrm{v}(a)\|_{L^\infty(\mathbb{R}^n)} \le C\|a\|_{l^\infty}$$

for some C independent of a and $\mathbf{m}$. Furthermore, analogously to Corollary 4.4, all Sobolev norms are also equivalent on $T_{\mathbf{m}}\mathbb{P}(L)$.

EXAMPLE 4.7. Let us consider the one-dimensional case $x \in \mathbb{R}^1$ and assume, in addition, that the symmetry group G consists of only shifts $G = \mathbb{R}$. Then, the point $\mathbf{m} \in \mathbb{P}(L)$ is determined by a sequence $\{\xi_i\}_{i\in\mathbb{Z}}$, $\xi_i \in \mathbb{R}$ of the pulse centers. Moreover, in the 1D case there is a natural ordering: $\xi_i > \xi_j$ for $i > j$ which allows to determine the point $\mathbf{m}$ in a unique way. To be more precise, the manifold $\mathbb{P}(L)$ has the following structure:

$$\mathbb{P}(L) := \{\mathbf{m} := \sum_{i=-\infty}^{\infty} V(x-\xi_i), \ \ \xi_i \in \mathbb{R}, \ \ \xi_{i+1} > \xi_i, \ \ \sup_{i\in\mathbb{Z}} |\xi_{i+1}-\xi_i| > 2L\}. \tag{4.28}$$

However, in the multi-dimensional case, there are not any reasonable order on the set of the pulse centers and we have to use the quotient (4.4).

We are now ready to prove that the manifold $\mathbb{P}(L)$ consists of "almost equilibria" of equation (2.1). To this end, we introduce the function

$$\mathbb{F}(\mathbf{m}) := A_0\mathbf{m} + \Phi(\mathbf{m}), \quad \mathbf{m} \in \mathbb{P}(L) \tag{4.29}$$

or, in terms of $\vec{\Gamma}$,

$$\mathbb{F}(\vec{\Gamma}) := \mathbb{F}(\mathbb{V}_{\vec{\Gamma}}) = \Phi(\mathbb{V}_{\vec{\Gamma}}) - \sum_{j=1}^{\infty} \Phi(V_{\Gamma_j}) \tag{4.30}$$

where $\vec{\Gamma} = [\mathbb{V}]^{-1}\mathbf{m}$ (here we have used that every pulse V_{Γ_j} is an equilibrium of (2.1)).

The following lemma shows that the function $\mathbb{F}$ is indeed small.

LEMMA 4.8. *Let the above assumptions hold. Then, for every $\varepsilon > 0$ the function $\mathbb{F}(\vec{\Gamma})$ satisfies*

$$|\mathbb{F}(\vec{\Gamma})(x)| \le C_\varepsilon\, e^{-(\alpha-\varepsilon)[\operatorname{dist}(x,\Xi)+\operatorname{dist}'(x,\Xi)]}, \tag{4.31}$$

where the positive constant C_ε is independent of $x \in \mathbb{R}^n$ and $\vec{\Gamma} \in \mathbb{B}(L)$ and $\operatorname{dist}'(x,\Xi)$ denotes the distance from x to the second nearest element of Ξ:

$$\operatorname{dist}'(x,\Xi) := \sup_{j\in\mathbb{N}} \inf_{i\neq j} \|x-\xi_i\|. \tag{4.32}$$

In particular, $\|\mathbb{F}(\vec{\Gamma})\|_{L^\infty(\mathbb{R}^n)} \le C_\varepsilon\, \mathrm{e}^{-2(\alpha-\varepsilon)L}$, $\varepsilon > 0$. *Moreover, the function* $\mathbb{F}$: $\vec{\Gamma} \to L^\infty(\mathbb{R}^n)$ *is Fréchet differentiable and*

$$\|\mathbb{F}'(\vec{\Gamma})\|_{\mathcal{L}(T_{\vec{\Gamma}}\mathbb{B}(L),L^\infty(\mathbb{R}^n))} \le C_\varepsilon\, \mathrm{e}^{-2(\alpha-\varepsilon)L}\,. \tag{4.33}$$

PROOF. In order to verify (4.31), we use the following obvious formula

$$\Phi(v_1+v_2)-\Phi(v_1)-\Phi(v_2) = \int_0^1 \Phi'(v_1+sv_2)\,\mathrm{d}s\cdot v_2- \\ -\int_0^1 \Phi'(sv_2)\,\mathrm{d}s\cdot v_2 = \int_0^1\int_0^1 \Phi''(s_1v_1+s_2v_2)\,\mathrm{d}s_1\,\mathrm{d}s_2\cdot[v_1,v_2] \tag{4.34}$$

which is valid for every two sufficiently smooth functions v_1 and v_2. Applying this formula to the function $\mathbb{F}(\vec{\Gamma})$ and iterating it (infinitely many times), we have

$$\mathbb{F}(\vec{\Gamma}) = \sum_{i=1}^\infty \mathbb{T}(V_{\Gamma_i}, \sum_{j=i+1}^\infty V_{\Gamma_j})[V_{\Gamma_i}, \sum_{j=i+1}^\infty V_{\Gamma_j}] \tag{4.35}$$

where $\mathbb{T}(u,v) := \int_0^1\int_0^1 D_u^2\Phi(s_1u+s_2v)\,\mathrm{d}s_1\,\mathrm{d}s_2$.

Using now estimates (2.28) and the fact that all of $\sum_{j=i+1}^\infty V_{\Gamma_j}$ are uniformly bounded, we obtain the following estimate:

$$|(\mathbb{F}(\vec{\Gamma}))(x)| \le C\sum_{i,j,\,i\ne j} \mathrm{e}^{-\alpha|x-\xi_i|}\,\mathrm{e}^{-\alpha|x-\xi_j|}\,. \tag{4.36}$$

This estimate implies (4.31). Indeed, let $\xi_{j_0}\in\Xi$ be the nearest to x element of the grid Ξ. Then, according to Lemma 3.12,

$$\sum_{i,j,\ i\ne j} \mathrm{e}^{-\alpha|x-\xi_i|}\,\mathrm{e}^{-\alpha|x-\xi_j|} = \mathrm{e}^{-\alpha|x-\xi_{j_0}|}\sum_{i\ne j_0}\mathrm{e}^{-\alpha|x-\xi_i|} + \sum_{j\ne j_0}\mathrm{e}^{-\alpha|x-\xi_j|}\sum_{i\ne j}\mathrm{e}^{-\alpha|x-\xi_i|} \\ \le C_\varepsilon\,\mathrm{e}^{-\alpha\operatorname{dist}(x,\Xi)}\,\mathrm{e}^{-(\alpha-\varepsilon)\operatorname{dist}(x,\Xi\backslash\{\xi_{j_0}\})} + \sum_{j\ne j_0}\mathrm{e}^{-\alpha|x-\xi_j|}\,C_\varepsilon\,\mathrm{e}^{-(\alpha-\varepsilon)\operatorname{dist}(x,\Xi)} \le \\ \le C_\varepsilon\,\mathrm{e}^{-(\alpha-\varepsilon)[\operatorname{dist}(x,\Xi)+\operatorname{dist}'(x,\Xi)]}\,. \tag{4.37}$$

Moreover, since

$$\operatorname{dist}(x,\Xi)+\operatorname{dist}'(x,\Xi)\ge|\xi_{j_0}-\xi_{j_0'}|\ge 2L$$

where $\xi_{j_0'}$ is the second nearest to x element of the grid Ξ, estimate (4.31) implies that the L^∞-norm of $\mathbb{F}(\vec{\Gamma})$ is bounded from above by $C_\varepsilon\,\mathrm{e}^{-2(\alpha-\varepsilon)L}$.

The Fréchet differentiability of $\mathbb{F}(\vec{\Gamma})$ can be verified in a standard way and we leave it for the reader. So, it only remains to verify estimate (4.33). To this end, we note that the derivative $D_{\vec{\Gamma}}\mathbb{F}(\vec{\Gamma})\delta\vec{\Gamma}$, $\delta\vec{\Gamma}\in T_{\vec{\Gamma}}\mathbb{B}(L)$ is given by the following expression:

$$\mathbb{F}'(\vec{\Gamma})\delta\vec{\Gamma} = \sum_{j=1}^\infty[\Phi'(\mathbb{V}_{\vec{\Gamma}})-\Phi'(V_{\Gamma_j})]D_\Gamma V_{\Gamma_j}\cdot\delta\Gamma_j. \tag{4.38}$$

Using now the obvious formula

$$\Phi'(\vec{\Gamma})-\Phi'(V_{\Gamma_j})) = \int_0^1\Phi''(\mathbb{V}_{\vec{\Gamma}}-sV_{\Gamma_j})\,\mathrm{d}s\cdot\sum_{i\ne j}V_{\Gamma_i}$$

together with estimates (2.28) and Lemma 3.12, we deduce analogously to the proof of (4.31) that

$$(4.39)\quad |(\mathbb{F}'(\vec{\Gamma})\delta\vec{\Gamma})(x)| \le \sum_{j=1}^{\infty} \mathrm{e}^{-\alpha|x-\xi_j|}\,\|\delta\Gamma_j\| \sum_{i\ne j} \mathrm{e}^{-\alpha|x-\xi_i|} \le$$

$$\le C\|\delta\vec{\Gamma}\|_{\mathbb{R}_\infty} \sum_{j=1}^{\infty} \mathrm{e}^{-\alpha|x-\xi_j|} \sum_{i\ne j} \mathrm{e}^{-\alpha|x-\xi_i|} \le C_\varepsilon \|\delta\vec{\Gamma}\|_{l^\infty}\, \mathrm{e}^{-(\alpha-\varepsilon)[\mathrm{dist}(x,\Xi)+\mathrm{dist}'(x,\Xi)]}$$

and Lemma 4.8 is proven. □

In particular, estimates (4.31) and (4.33) imply the following uniform Lipschitz continuity of the function $\mathbb{F}$:

$$(4.40)\qquad \|\mathbb{F}(\vec{\Gamma}^1) - \mathbb{F}(\vec{\Gamma}^2)\|_{L^\infty(\mathbb{R}^n)} \le C_\varepsilon\, \mathrm{e}^{-2(\alpha-\varepsilon)L}\, d^s(\vec{\Gamma}^1, \vec{\Gamma}^2)$$

where the constant C_ε depends on $\varepsilon > 0$, but is independent of $\vec{\Gamma}^1, \vec{\Gamma}^2 \in \mathbb{B}(L)$. The next lemma gives the weighted analog of this estimate.

LEMMA 4.9. *Let the above assumptions hold and $\gamma \le \alpha/2$. Then*

$$(4.41)\qquad \|\mathbb{F}(\vec{\Gamma}^1) - \mathbb{F}(\vec{\Gamma}^2)\|_{L^\infty_{\mathrm{e}^{-\gamma|x|}}(\mathbb{R}^n)} \le C\,\mathrm{e}^{-\alpha L}\, d^s_\gamma(\vec{\Gamma}^1, \vec{\Gamma}^2)$$

where the constant C is independent of $\vec{\Gamma}^i \in \mathbb{B}(L)$.

PROOF. Let $\vec{\Gamma}^i \in \mathbb{B}(L)$, $i = 1, 2$. Analogously to Proposition 4.3 We introduce the sets $\mathbb{N}^i_o = \mathbb{N}^i_o(1)$ and $\mathbb{N}_c(r) = \mathbb{N}_c(1)$ via (4.13), (4.14) and will use splitting (4.16). Then, we have

$$(4.42)\quad |[\mathbb{F}(\vec{\Gamma}^1) - \mathbb{F}(\vec{\Gamma}^2)](x)| \le |[\mathbb{F}(\vec{\Gamma}^1) - \mathbb{F}(\vec{\Gamma}^1_c)](x)| +$$

$$+ |[\mathbb{F}(\vec{\Gamma}^2) - \mathbb{F}(\vec{\Gamma}^2_c)](x)| + |[\mathbb{F}(\vec{\Gamma}^1_c) - \mathbb{F}(\vec{\Gamma}^2_c)](x)|.$$

We transform the first term in the right-hand side of (4.42) as follows:

$$|[\mathbb{F}(\vec{\Gamma}^1) - \mathbb{F}(\vec{\Gamma}^1_c)](x)| \le |[\Phi(\mathbb{V}_{\vec{\Gamma}^1_c} + \mathbb{V}_{\vec{\Gamma}^1_o}) - \Phi(\mathbb{V}_{\vec{\Gamma}^1_c}) - \Phi(\mathbb{V}_{\vec{\Gamma}^1_o})](x)| + |\mathbb{F}(\vec{\Gamma}^1_o)(x)|.$$

Using now (4.34) and Lemma 3.12, we infer

$$|[\Phi(\mathbb{V}_{\vec{\Gamma}^1_c} + \mathbb{V}_{\vec{\Gamma}^1_o}) - \Phi(\mathbb{V}_{\vec{\Gamma}^1_c}) - \Phi(\mathbb{V}_{\vec{\Gamma}^1_o})](x)| \le \sum_{i\in\mathbb{N}^1_c}\sum_{j\in\mathbb{N}^1_o} \mathrm{e}^{-\alpha|x-\xi^1_i|-\alpha|x-\xi^1_j|} \le$$

$$\le C \sup_{j\in\mathbb{N}^1_o}\{\mathrm{e}^{-\alpha|x-\xi^1_j|/4}\}\,\mathrm{e}^{-\alpha/2(\mathrm{dist}(x,\Sigma^1_c)+\mathrm{dist}(x,\Sigma^1_o))} \le c\,\mathrm{e}^{-\alpha L} \sup_{j\in\mathbb{N}^1_o}\{\mathrm{e}^{-\alpha|x-\xi^1_j|/4}\}.$$

Multiplying this estimate by $\mathrm{e}^{-\gamma|x|}$ and using that $\mathrm{e}^{-\gamma|x|-\gamma|x-\xi_j|} \le \mathrm{e}^{-\gamma|\xi_j|}$ and $\gamma \le \alpha/4$, we have

$$\|\Phi(\mathbb{V}_{\vec{\Gamma}^1_c} + \mathbb{V}_{\vec{\Gamma}^1_o}) - \Phi(\mathbb{V}_{\vec{\Gamma}^1_c}) - \Phi(\mathbb{V}_{\vec{\Gamma}^1_o})\|_{L^\infty_{\mathrm{e}^{-\gamma|x|}}(\mathbb{R}^n)} \le C \sup_{j\in\mathbb{N}^1_o}\{\mathrm{e}^{-\gamma|\xi^1_j|}\}.$$

Moreover, using (4.36) (with $\vec{\Gamma}^1_o$ instead of $\vec{\Gamma}$), we deduce the analogous estimate for the term $\mathbb{F}(\vec{\Gamma}^1_o)$. Thus,

$$(4.43)\qquad \|\mathbb{F}(\vec{\Gamma}^1) - \mathbb{F}(\vec{\Gamma}^1_c)\|_{L^\infty_{\mathrm{e}^{-\gamma|x|}}(\mathbb{R}^n)} \le CD^s_\gamma(\vec{\Gamma}^1, \vec{\Gamma}^2)$$

Analogously, the second term in the right-hand side of (4.42) can be estimated via

$$(4.44)\qquad \|\mathbb{F}(\vec{\Gamma}^2) - \mathbb{F}(\vec{\Gamma}^2_c)\|_{L^\infty_{\mathrm{e}^{-\gamma|x|}}(\mathbb{R}^n)} \le CD^s_\gamma(\vec{\Gamma}^1, \vec{\Gamma}^2).$$

In order to estimate the third term, we recall that $\|\vec{\Gamma}^1_c - \vec{\Gamma}^2_c\|_\infty \le 1$ and, consequently, analogously to (4.39), we have

$$|[\mathbb{F}(\vec{\Gamma}^1_c) - \mathbb{F}(\vec{\Gamma}^2_c)](x)| \le C_\varepsilon \, \mathrm{e}^{-\alpha L} \sup_{j\in\mathbb{N}_c} \{\mathrm{e}^{-\alpha|x-\xi^1_j|/4} \, \|\Gamma^1_j - \Gamma^2_j\|\}.$$

Multiplying this estimate by $\mathrm{e}^{-\gamma|x|}$ and taking the supremum over x, we finally deduce that

$$\|\mathbb{F}(\vec{\Gamma}^1_c) - \mathbb{F}(\vec{\Gamma}^2_c)\|_{L^\infty_{\mathrm{e}^{-\gamma|x|}}(\mathbb{R}^n)} \le C \, \mathrm{e}^{-\alpha L} \, D^s_\gamma(\vec{\Gamma}^1, \vec{\Gamma}^2). \tag{4.45}$$

Estimates (4.43), (4.44) and (4.45) imply (4.41) and finish the proof of Lemma 4.9. □

We now reformulate the above two lemmas in terms of the manifold $\mathbb{P}(L)$.

COROLLARY 4.10. *Let the above assumptions hold. Then the function* $\mathbb{F} : \mathbb{P}(L) \to L^\infty(\mathbb{R}^n)$ *is Fréchet differentiable and satisfies the following estimates:*

$$\begin{aligned} &\|\mathbb{F}(\mathbf{m}_1) - \mathbb{F}(\mathbf{m}_2)\|_{L^\infty(\mathbb{R}^n)} \le C_\varepsilon \, \mathrm{e}^{-2(\alpha-\varepsilon)L} \, \|\mathbf{m}_1 - \mathbf{m}_2\|_{L^\infty(\mathbb{R}^n)} \\ &\|\mathbb{F}'(\mathbf{m})\|_{\mathcal{L}(T_{\mathbf{m}}\mathbb{P}(L), L^\infty(\mathbb{R}^n))} \le C_\varepsilon \, \mathrm{e}^{-2(\alpha-\varepsilon)L} \end{aligned} \tag{4.46}$$

where $\varepsilon > 0$ *is arbitrary and* C_ε *depends on* ε*, but is independent of* L *and* $\mathbf{m}_i \in \mathbb{P}(L)$ *(here and below we denote by* $T_{\mathbf{m}}\mathbb{P}(L)$ *the tangent space to* $\mathbb{P}(L)$ *at point* $\mathbf{m}$*).*

Moreover, for every $0 \le \gamma \le \alpha/4$*, we have*

$$\|\mathbb{F}(\mathbf{m}_1) - \mathbb{F}(\mathbf{m}_2)\|_{L^\infty_{\mathrm{e}^{-\gamma|x|}}(\mathbb{R}^n)} \le C \, \mathrm{e}^{-\alpha L} \, \|\mathbf{m}_1 - \mathbf{m}_2\|_{L^\infty_{\mathrm{e}^{-\gamma|x|}}(\mathbb{R}^n)} \tag{4.47}$$

where C *is independent of* L *and* $\mathbf{m}_i \in \mathbb{P}(L)$.

Indeed, the assertion of this corollary follows immediately from Lemma 4.8 and 4.9 and Proposition 4.3.

REMARK 4.11. Arguing analogously, it is not difficult to check that the map $\mathbb{F} \in C^k(\mathbb{P}(L), L^\infty(\mathbb{R}^n))$ for every $k \in \mathbb{N}$ and the analogies of estimates (4.46) hold. Moreover, the weight $\mathrm{e}^{-\gamma|x|}$ in Proposition 4.3 and Lemma 4.9 can be replaced by $\mathrm{e}^{-\gamma|x-x_0|}$ and all of the constants will be uniform with respect to $x_0 \in \mathbb{R}^n$. This gives estimate (4.47) also with weight $\mathrm{e}^{-\gamma|x-x_0|}$. Proposition 3.3 then gives estimate (4.47) for any weight function θ of sufficiently small exponential growth rate.

5. The multi-pulse manifold: projectors and tangent spaces

In this section, we continue to study the basic manifold $\mathbb{P}(L)$. To be more precise, we construct here a family of projectors $\mathbb{P}_{\mathbf{m}} : [L^\infty(\mathbb{R}^n)]^m \to T_{\mathbf{m}}\mathbb{P}(L)$ which plays an essential role in our center manifold construction. To this end, we need the following lemmas.

LEMMA 5.1. *Let the assumptions of Section 2 hold. Then, for sufficiently large* L *and every* $\vec{\Gamma} \in \mathbb{B}(L)$ *there exists a family of functions* $\bar{\psi}^i_j(\vec{\Gamma}) \in C^\infty(\mathbb{R}^n)$, $i = 1, \cdots, k$, $j \in \mathbb{N}$ *which satisfy the analog of* (2.32)*:*

$$|\partial^p_x \bar{\psi}^i_j(\vec{\Gamma})| \le C_{p,\varepsilon} \, \mathrm{e}^{-(\alpha-\varepsilon)|x-\xi_j|} \tag{5.1}$$

and the following orthogonality relations

$$(\bar{\psi}^i_j(\vec{\Gamma}), \phi^l_{\Gamma_r}) = \delta_{il} \cdot \delta_{jr}. \tag{5.2}$$

Moreover, these functions are Fréchet differentiable with respect to $\vec{\Gamma}$ *and the following uniform estimate holds:*

$$|D_{\vec{\Gamma}}\bar{\psi}^i_j(\vec{\Gamma})(x)\delta\vec{\Gamma}| \le C_\varepsilon\, e^{-(\alpha-\varepsilon)|x-\xi_j|}\, \|\delta\vec{\Gamma}\|_{l^\infty} \tag{5.3}$$

where the constant C_ε *depends on* $\varepsilon > 0$*, but is independent of* $\vec{\Gamma} \in \mathbb{B}(L)$*,* i*,* j *and* $x \in \mathbb{R}^n$*.*

PROOF. Let us fix $j_0 \in \mathbb{N}$ and seek for the associated functions $\bar{\psi}^i_{j_0}$ in the form

$$\bar{\psi}^i_{j_0}(x) = \sum_{j=1}^{\infty}\sum_{i=1}^{k} a^i_j \psi^i_{\Gamma_j}(x) \tag{5.4}$$

where $a^i_j \in \mathbb{R}$ are the unknown coefficients. Multiplying (5.4) scalarly by $\phi^l_{\Gamma_r}$ and using the orthogonality relations (2.35) and (5.2), we obtain the following system for a^i_j:

$$a^l_r + \sum_{j\ne r}\sum_{i=1}^{k} a^i_j(\psi^i_{\Gamma_j}, \phi^l_{\Gamma_r}) = \delta_{j_0 r}. \tag{5.5}$$

We claim that, for sufficiently large L this equation in l^∞ possesses a unique solution and this solution possesses the following estimate:

$$|a^i_j| \le 2\, e^{-(\alpha-\varepsilon)|\xi_j - \xi_{j_0}|}. \tag{5.6}$$

Indeed, let us introduce a new unknown $b^i_j := a^i_j\, e^{(\alpha-2\varepsilon)|\xi_j-\xi_{j_0}|}$ which satisfies the following analog of (5.5):

$$\vec{b} + \mathbb{B}_{j_0}\vec{b} = \vec{l}_{j_0} \tag{5.7}$$

where $(\vec{l}_{j_0})^i_j := \delta_{j,j_0}$ and the linear operator $\mathbb{B}_{j_0}$ is defined via

$$(\mathbb{B}_{j_0}\vec{l})^l_r := b^l_r + \sum_{j\ne r}\sum_{i=1}^{k} b^i_j\, e^{-(\alpha-2\varepsilon)(|\xi_j-\xi_{j_0}|-|\xi_r-\xi_{j_0}|)}(\psi^i_{\Gamma_j}, \phi^l_{\Gamma_r}). \tag{5.8}$$

We now note that, due to estimates (2.32), we have

$$\begin{aligned} & e^{-(\alpha-2\varepsilon)(|\xi_j-\xi_{j_0}|-|\xi_r-\xi_{j_0}|)}\, |(\psi^i_{\Gamma_j}, \phi^l_{\Gamma_r})| \le \\ & \qquad \le C_\varepsilon\, e^{(\alpha-2\varepsilon)|\xi_r-\xi_j|}\, e^{-(\alpha-\varepsilon)|\xi_r-\xi_j|} = C_\varepsilon\, e^{-\varepsilon|\xi_r-\xi_j|} \end{aligned} \tag{5.9}$$

and, consequently, due to Lemma 3.12,

$$\|\mathbb{B}_{j_0}\|_{\mathcal{L}(l^\infty, l^\infty)} \le C'_\varepsilon L^n\, e^{-\varepsilon L}. \tag{5.10}$$

Thus, for sufficiently large L_0, $\|\mathbb{B}_{j_0}\| \le 1/2$ and, consequently, equation (5.7) is uniquely solvable in l^∞ and the solution $\vec{b}$ satisfies $\|\vec{b}\|_\infty \le 2\|\vec{l}_{j_0}\|_\infty = 2$ which implies estimate (5.6).

It remains to note that (5.6) together with (2.32) imply (5.1). Indeed,

$$\begin{aligned} |\partial^p_x \bar{\psi}^i_j(\vec{\Gamma})(x)| & \le 2C_p \sum_{r=1}^{\infty} e^{-(\alpha-\varepsilon)|\xi_r-\xi_j|}\, e^{-\alpha|x-\xi_r|} \le \\ & \le 2C_p\, e^{-(\alpha-\varepsilon)|x-\xi_j|} \sum_{r=1}^{\infty} e^{-\varepsilon|x-\xi_r|} \le C'_{\varepsilon,p}\, e^{-(\alpha-\varepsilon)|x-\xi_j|}. \end{aligned} \tag{5.11}$$

Thus, (5.1) is verified. The differentiability of the functions $\bar\psi_j^i$ with respect to $\vec\Gamma$ and estimate (5.3) can be verified analogously (see also the next Lemma). Lemma 5.1 is proven. $\square$

We are now ready to define the required projectors $\mathbb{P}_{\mathbf{m}}$:

$$\mathbb{P}_{\mathbf{m}}v := \sum_{j\in\mathbb{N}}\sum_{i=1}^{k}(\bar\psi_j^i(\vec\Gamma), v)\phi_{\Gamma_j}^i \tag{5.12}$$

where $\mathbf{m} \in \mathbb{P}(L)$, $\vec\Gamma = [\mathbb{V}]^{-1}\mathbf{m}$ and $v \in L^\infty(\mathbb{R}^n)$ is arbitrary. Indeed, it follows immediately from (5.12) and the fact that ϕ_{Γ_j} is a derivative of V_{Γ_j} with respect to Γ_j, that

$$\mathbb{P}_{\mathbf{m}}v \subset T_{\mathbf{m}}\mathbb{P}(L), \quad \forall v \in L^\infty(\mathbb{R}^n).$$

Moreover, from the orthogonality relations (5.2), we infer

$$\mathbb{P}_{\mathbf{m}}^2 = \mathbb{P}_{\mathbf{m}}, \quad \mathbb{P}_{\mathbf{m}}T_{\mathbf{m}}\mathbb{P}(L) = T_{\mathbf{m}}\mathbb{P}(L). \tag{5.13}$$

Thus, $\mathbb{P}_{\mathbf{m}}$ are indeed the projectors on the tangent space $T_{\mathbf{m}}\mathbb{P}(L)$.

The following theorem which establishes the smooth dependence of the projectors $\mathbb{P}_{\mathbf{m}}$ on $\mathbf{m}$ can be considered as the main result of the section.

THEOREM 5.2. *Let the above assumptions hold. Then, the projectors $\mathbb{P}_{\mathbf{m}}$ are well-defined and depend smoothly on $\mathbf{m} \in \mathbb{P}(L)$. In particular, for every $1 \le p \le \infty$,*

$$\|\mathbb{P}_{\mathbf{m}}\|_{\mathcal{L}(L^p(\mathbb{R}^n),L^p(\mathbb{R}^n))} + \|\mathbb{P}'_{\mathbf{m}}\|_{\mathcal{L}(T_{\mathbf{m}}\mathbb{P}(L),\mathcal{L}(L^p(\mathbb{R}^n),L^p(\mathbb{R}^n)))} \le C \tag{5.14}$$

where the constant C is independent of $\mathbf{m}$, p and L (if L is large enough). Moreover, for $v \in L^1_{\mathrm{b}}(\mathbb{R}^n)$ and sufficiently small $\gamma > 0$, we also have

$$\|\mathbb{P}_{\mathbf{m}_1}v - \mathbb{P}_{\mathbf{m}_2}v\|_{L^\infty_{e^{-\gamma|x|}}(\mathbb{R}^n)} \le C\|v\|_{L^1_{\mathrm{b}}(\mathbb{R}^n)}\|\mathbf{m}_1 - \mathbf{m}_2\|_{L^\infty_{e^{-\gamma|x|}}(\mathbb{R}^n)} \tag{5.15}$$

uniformly with respect to $\mathbf{m}_1, \mathbf{m}_2 \in \mathbb{P}(L)$.

PROOF. Indeed, estimate (5.14) is a standard corollary of Lemma 5.1, Proposition 3.13 and exponential decay conditions (2.32), so we rest its proof to the reader.

Thus, we only need to verify the weighted Lipschitz continuity (5.15). Indeed, let $\Gamma^1, \Gamma^2 \in \mathbb{B}(L)$ be arbitrary and let the sets $\mathbb{N}_o^i$, $i = 1, 2$ and $\mathbb{N}_c$ be defined by (5.61) and (5.62) with $r = 1$. Then, we split the difference $\mathbb{P}_{\mathbf{m}_1}v - \mathbb{P}_{\mathbf{m}_2}v$ as follows:

$$\begin{aligned}\mathbb{P}_{\mathbf{m}_1}v - \mathbb{P}_{\mathbf{m}_2}v = &\sum_{j\in\mathbb{N}_o^1}\sum_{i=1}^{k}(\bar\psi_j^i(\vec\Gamma^1), v)\phi_{\Gamma_j^1}^i - \sum_{j\in\mathbb{N}_o^2}\sum_{i=1}^{k}(\bar\psi_j^i(\vec\Gamma^2), v)\phi_{\Gamma_j^2}^i +\\ &+ \sum_{j\in\mathbb{N}_c}\sum_{i=1}^{k}(\bar\psi_j^i(\vec\Gamma^1), v)[\phi_{\Gamma_j^1}^i - \phi_{\Gamma_j^2}^i] - \sum_{j\in\mathbb{N}_c}\sum_{i=1}^{k}(\bar\psi_j^i(\vec\Gamma^1) - \bar\psi_j^i(\vec\Gamma^2), v)\phi_{\Gamma_j^2}^i\end{aligned} \tag{5.16}$$

Using now inequalities (4.10), (2.32), we estimate the first term in the right-hand side of (5.16) via

$$\Big|\sum_{j\in\mathbb{N}_o^1}\sum_{i=1}^{k}(\bar\psi_j^i(\vec\Gamma^1), v)\phi_{\Gamma_j^1}^i\Big| \le C\|v\|_{L^1_{\mathrm{b}}(\mathbb{R}^n)}\sum_{j\in\mathbb{N}_o^1} e^{-\alpha|x-\xi_j^1|}$$

which immediately gives

$$\|\sum_{j\in\mathbb{N}_o^1}\sum_{i=1}^{k}(\bar{\psi}_j^i(\vec{\Gamma}^1),v)\phi^i_{\Gamma_j^1}\|_{L^\infty_{e^{-\gamma|x|}}(\mathbb{R}^n)}\le C\sup_{j\in\mathbb{N}_0^1}\{e^{-\gamma|\xi_j^1|}\}$$

if $\gamma<\alpha/2$. The second and the third term in (5.16) can be estimated analogously. And, the last forth term can be also estimated analogously if, in addition, the following estimate is known:

$$\|\bar{\psi}_j^i(\vec{\Gamma}^1)-\bar{\psi}_j^i(\vec{\Gamma}^2)\|_{L^\infty_{e^{-\gamma|\xi_j^1|+\alpha|x-\xi_j^1|/2}}(\mathbb{R}^n)}\le CD^s_\gamma(\vec{\Gamma}^1,\vec{\Gamma}^2) \tag{5.17}$$

for every $j\in\mathbb{N}_c$ (we recall that the metric D^s_γ is introduced by (4.15)). In order to verify this, we recall that the function $\psi^i_{j_0}(\vec{\Gamma})$ satisfy

$$\bar{\psi}_{j_0}(\vec{\Gamma})=\sum_{j=1}^{\infty}a_j(j_0,\vec{\Gamma})\psi_{\Gamma_j}$$

(in order to simplify the notations, we assume that $k=1$ and omit the corresponding indexes). Then, using (5.6) and (2.32) and arguing analogously, we reduce the verification of (5.17) to the following estimate:

$$|a_j(j_0,\vec{\Gamma}^1)-a_j(j_0,\vec{\Gamma}^2)|\,e^{-\gamma|\xi_j^1|+\alpha|\xi_j^1-\xi_{j_0}^1|/2}\le CD^s_\gamma(\vec{\Gamma}^1,\vec{\Gamma}^2) \tag{5.18}$$

for all $j,j_0\in\mathbb{N}_c$ and with C independent of j, j_0 and $\vec{\Gamma}^i$.

In order to verify (5.18), we recall that the coefficients $a_j(j_0,\vec{\Gamma})$ are determined by the following equations:

$$a_r(j_0,\vec{\Gamma})+\sum_{j=1}^{\infty}A_r^j(\vec{\Gamma})a_j(j_0,\vec{\Gamma})=\delta_{r,j_0},\quad r\in\mathbb{N}$$

with $A_r^j(\vec{\Gamma})=(\psi_{\Gamma_r},\phi_{\Gamma_j})$. Let us introduce now the new coefficients:

$$c_r(j_0,\vec{\Gamma}):=e^{-\gamma|\xi_r|}\,e^{\alpha|\xi_{j_0}-\xi_r|/2}\,a_r(j_0,\vec{\Gamma}),\quad d_r(j_0):=c_r(j_0,\vec{\Gamma}^1)-c_r(j_0,\vec{\Gamma}^2).$$

Then, the coefficients c_r solve

$$c_r(j_0,\vec{\Gamma})+\sum_{j\neq r}C_r^j(j_0,\vec{\Gamma})c_j(j_0,\vec{\Gamma})=e^{-\gamma|\xi_r|}\,\delta_{r,j}$$

with $C_r^j(j_0,\Gamma):=e^{-\gamma|\xi_r|}\,e^{\gamma|\xi_j|}\,e^{\alpha|\xi_{j_0}-\xi_r|/2}\,e^{-\alpha|\xi_{j_0}-\xi_j|/2}\,A_r^j(j_0,\vec{\Gamma})$ and, moreover, if $\gamma<\alpha/4$, we also have

$$|C_r^j(j_0,\vec{\Gamma})|\le C\,e^{-\alpha|\xi_j-\xi_r|/4}\,. \tag{5.19}$$

Furthermore, we rewrite the equations for $\{d_r\}_{r\in\mathbb{N}_c}$ as follows:

$$d_r(j_0)+\sum_{j\in\mathbb{N}_c,\,j\neq r}C_r^j(j_0,\vec{\Gamma}^1)d_j(j_0)=L_r(j_0) \tag{5.20}$$

where $r,j\in\mathbb{N}_c$ and

$$\begin{aligned}L_r(j_0):=(e^{-\gamma|\xi_r^1|}-e^{-\gamma|\xi_r^2|})\delta_{r,j_0}-\sum_{j\in\mathbb{N}_o^1}C_r^j(j_0,\vec{\Gamma}^1)c_j(j_0,\vec{\Gamma}^1)+\\+\sum_{j\in\mathbb{N}_o^2}C_r^j(j_0,\vec{\Gamma}^2)c_j(j_0,\vec{\Gamma}^2)+\sum_{j\in\mathbb{N}_c}[C_r^j(j_0,\vec{\Gamma}^1)-C_r^j(j_0,\vec{\Gamma}^2)]c_j(j_0,\vec{\Gamma}^2).\end{aligned} \tag{5.21}$$

We claim that

$$|L_r(j_0)| \le CD^s_\gamma(\vec\Gamma^1,\vec\Gamma^2) \tag{5.22}$$

where the constant C is independent of $r, j_0 \in \mathbb{N}_c$ and $\vec\Gamma^i \in \mathbb{B}(L)$.

Indeed, since $r \in \mathbb{N}_c$, then estimate (5.22) for the first term in the right-hand side of (5.21) is obvious. Analogously, using (5.6) and (5.19) together with Lemma 3.12, we infer

$$|\sum_{j\in\mathbb{N}^1_o} C^j_r(j_0,\vec\Gamma^1)c_j(j_0,\vec\Gamma^1)| \le$$
$$\le C\sum_{j\in\mathbb{N}^1_o} e^{-\alpha|\xi^1_j-\xi^1_r|}\, e^{-\gamma|\xi^1_j|}\, e^{-\alpha|\xi_j-\xi_{j_0}|/4} \le C_1 \sup_{j\in\mathbb{N}^1_o}\{e^{-\gamma|\xi_j|}\}.$$

The third term is completely analogous and we only need to estimate the last one. To this end, we use again the fact that $j, j_0, r \in \mathbb{N}_c$, estimates (5.6) and (5.19) and, in addition, estimates (2.38) (for estimating the differences between $\psi_{\Gamma^1_j}$ and $\psi_{\Gamma^2_j}$, see the above formula for $C^j_r(j_0,\vec\Gamma)$). Then, we obtain

$$|\sum_{j\in\mathbb{N}_c}[C^j_r(j_0,\vec\Gamma^1) - C^j_r(j_0,\vec\Gamma^2)]c_j(j_0,\vec\Gamma^2)| \le$$
$$\le C\sum_{j\in\mathbb{N}_c} e^{-\alpha|\xi^1_j-\xi^1_r|/4}\, e^{-\gamma|\xi^1_j|}\, e^{-\alpha|\xi^1_{j_0}-\xi^1_j|/4}(\|\Gamma^1_j-\Gamma^2_j\| + \|\Gamma^1_{j_0}-\Gamma^2_{j_0}\| + \|\Gamma^1_r-\Gamma^2_r\|) \le$$
$$\le C\sup_{j\in\mathbb{N}_c}\{e^{-\gamma|\xi^1_j|}\,\|\Gamma^1_j-\Gamma^2_j|\}$$

(if $\gamma < \alpha/8$). Thus, estimate (5.22) is verified.

We are now able to return to system (5.20). Indeed, according to (5.19) and assumption (3.38), the matrix $\mathbb{C}(j_0,\vec\Gamma^1) := \{C^j_r(j_0,\vec\Gamma^1)\}_{j,r\in\mathbb{N}_c}$ is close to the diagonal one if L is large. Thus, fixing L to be large enough, analogously to the proof of Lemma 5.1, we infer

$$\sup_{j\in\mathbb{N}_c} d_j(j_0) \le 2\sup_{j\in\mathbb{N}_c} L_j(j_0) \le 2CD^s_\gamma(\vec\Gamma^1,\vec\Gamma^2)$$

which, together with (5.19) imply (5.18). Thus, estimate (5.15) is verified and Theorem 5.2 is proven. □

We are now formulate the analog of Lemma 3.1 and Corollary 3.4 for the projectors $\mathbb{P}_\mathbf{m}$.

PROPOSITION 5.3. *Let the above assumptions hold. Then, for sufficiently large L, there exists an exponent $\gamma > 0$ (independent of $L \ge L_0$) such that, for every $\mathbf{m} \in \mathbb{P}(L)$, every $1 \le p \le \infty$ and every weight function θ of exponential growth rate less than γ, one has*

$$\|\mathbb{P}_\mathbf{m} v\|_{\mathcal{L}(L^p_\theta(\mathbb{R}^n),L^p_\theta(\mathbb{R}^n))} + \|\mathbb{P}'_\mathbf{m}\|_{\mathcal{L}(T_\mathbf{m}\mathbb{P}(L),\mathcal{L}(L^p_\theta(\mathbb{R}^n),L^p_\theta(\mathbb{R}^n)))} \le C \tag{5.23}$$

where the constant C depends on C_θ, but is independent of $\mathbf{m}$, p and on the concrete choice of the weight θ.

Moreover, if M_{ε,x_0} be the multiplication operator on the special weight θ_{ε,x_0} and ε is small enough, then, analogously to (3.28),

$$\|M_{\varepsilon,x_0}\circ\mathbb{P}_\mathbf{m}\circ M_{\varepsilon,x_0} - \mathbb{P}_\mathbf{m}\|_{\mathcal{L}(L^p(\mathbb{R}^n),L^p(\mathbb{R}^n))} \le C|\varepsilon| \tag{5.24}$$

where the constant C is independent of ε, $\mathbf{m}$ and x_0.

The proof of that proposition is based on estimates (5.1) and (2.32) and is analogous to Corollary 3.14, see also Lemma 3.8, therefore, we leave it to the reader.

The next corollary gives the leading part of the asymptotics of $\mathbb{P}_{\mathbf{m}}$ as $L \to \infty$.

COROLLARY 5.4. *Let the above assumptions hold. Then, for sufficiently large L, every weight function of θ sufficiently small exponential growth rate and every $\mathbf{m} = \mathbb{V}_{\vec{\Gamma}} \in \mathbb{P}(L)$,*

$$\|\mathbb{P}_{\mathbf{m}} v - \sum_{j=1}^{\infty} \mathbb{P}_{\Gamma_j} v\|_{L^p_\theta(\mathbb{R}^n)} \le C_\varepsilon\, \mathrm{e}^{-2(\alpha-\varepsilon)L} \|v\|_{L^p_\theta(\mathbb{R}^n)} \tag{5.25}$$

where $\varepsilon > 0$ and $v \in L^p_\theta(\mathbb{R}^n)$ are arbitrary and C_ε depends on ε and C_θ, but is independent of $\mathbf{m}$ and v. Moreover, the analogous estimate holds for the spaces $L^p_{b,\theta}(\mathbb{R}^n)$ as well.

Indeed, according to (5.4), (5.6) and (2.32), we have

$$|\bar{\psi}^i_j(\vec{\Gamma})(x) - \psi^i_{\Gamma_j}(x)| \le C_\varepsilon\, \mathrm{e}^{-2(\alpha-\varepsilon)L}\, \mathrm{e}^{-\varepsilon|x-\xi_j|} \tag{5.26}$$

which together with definition (5.12) of the operator $\mathbb{P}_{\mathbf{m}}$ gives the required estimate (5.25).

EXAMPLE 5.5. Let us consider the self-adjoint 1D case without rotations ($G_0 = \mathbb{R}^1$ and consists only of translations). In particular, these simplifying assumptions will are satisfied for the Swift-Hohenberg equation, see Example 10.6 below. Then, the conjugate eigenfunction $\psi(x)$ coincide with $\phi(x) := V'(x)$ up to the renormalization:

$$\psi(x) = \frac{1}{\|\phi\|^2_{L^2}} \phi(x).$$

Furthermore, let us consider the case of only two pulses. Then, $\vec{\Gamma} := \xi := (\xi_1, \xi_2) \in \mathbb{R}^2$ with $\xi_2 - \xi_1 > 2L$ and the two-pulse configuration

$$\mathbf{m} = \mathbf{m}(\xi_1, \xi_2) := V(x - \xi_1) + V(x - \xi_2)$$

and, according to Lemma 5.1, the functions $\bar{\psi}_1 = \bar{\psi}_1(\xi_1, \xi_2)$ and $\bar{\psi}^2 \bar{\psi}_2(\xi_1, \xi_2)$ possess the following representation

$$\bar{\psi}_1(x) := a\phi(x - \xi_1) + b\phi(x - \xi_2), \quad \bar{\psi}_2(x) := c\phi(x - \xi_1) + d\phi(x - \xi_2)$$

where, according to the orthogonality conditions (5.4), the coefficients a, b, c, d can be found from the following linear system

$$\begin{cases} a\|\phi\|^2_{L^2} + bM(\xi) = 1, & aM(\xi) + b\|\phi\|^2_{L^2} = 0, \\ c\|\phi\|^2_{L^2} + dM(\xi) = 0, & cM(\xi) + d\|\phi\|^2_{L^2} = 1 \end{cases} \tag{5.27}$$

where $M(\xi) := (\phi(\cdot - \xi_1), \phi(\cdot - \xi_2))_{L^2}$. Thus, solving this system, we see that

$$a = d = \frac{\|\phi\|^2_{L^2}}{\|\phi\|^4_{L^2} - M^2(\xi)}, \quad b = c = -\frac{M(\xi)}{\|\phi\|^4_{L^2} - M^2(\xi)}$$

and, consequently,

$$\begin{cases} \bar{\psi}_1(x) = \frac{\|\phi\|^2_{L^2}}{\|\phi\|^4_{L^2} - M^2(\xi)} \phi(x - \xi_1) - \frac{M(\xi)}{\|\phi\|^4_{L^2} - M^2(\xi)} \phi(x - \xi_2) \\ \bar{\psi}_2(x) = -\frac{M(\xi)}{\|\phi\|^4_{L^2} - M^2(\xi)} \phi(x - \xi_1) + \frac{\|\phi\|^2_{L^2}}{\|\phi\|^4_{L^2} - M^2(\xi)} \phi(x - \xi_2). \end{cases} \tag{5.28}$$

Therefore, since, due to the exponential decaying condition for $\phi(x)$,

$$|M(\xi)| \le C\,e^{-(\alpha-\varepsilon)|\xi_2-\xi_1|} \le C\,e^{-2(\alpha-\varepsilon)L},$$

we see that the functions $\bar\psi_1$ and $\bar\psi_2$ are indeed well defined and close to $\psi(x-\xi_1)$ and $\psi(x-\xi_2)$ respectively if L is large enough (in accordance with the proven Lemma 5.1).

We introduce now one more linear operator which is essential for studying the evolution of pulses. We recall that $\mathbb{P}_{\mathbf{m}}$ is smooth with respect to $\mathbf{m}$, so the map $\delta\mathbf{m} \to \mathbb{P}'_{\mathbf{m}}[\delta\mathbf{m}]$ is well-defined for all $\delta\mathbf{m} \in T_{\mathbf{m}}\mathbb{P}(L)$. Then, we define

$$\mathbb{D}(\mathbf{m})[\delta\mathbf{m}]v := \mathbb{P}_{\mathbf{m}}\mathbb{P}'_{\mathbf{m}}[\delta\mathbf{m}]v \tag{5.29}$$

for every $\mathbf{m} \in \mathbb{P}(L)$, $\delta\mathbf{m} \in T_{\mathbf{m}}\mathbb{P}(L)$ and $v \in L^p_{\mathrm{b}}(\mathbb{R}^n)$ and obtain the following analog of Theorem 5.2.

THEOREM 5.6. *Let the above assumptions hold. Then, the operator $\mathbb{D}(\mathbf{m})$ is uniformly (with respect to $\mathbf{m} \in \mathbb{P}(L)$ and $L \ge L_0$) smooth (and, obviously, linear with respect to $\delta\mathbf{m}$ and v). Moreover, for sufficiently small γ, the following analog of* (5.15) *holds:*

$$\begin{aligned} &\|\mathbb{D}(\mathbf{m}_1)[\delta\mathbf{m}_1]v_1 - \mathbb{D}(\mathbf{m}_2)[\delta\mathbf{m}_2]v_2\|_{L^\infty_{e^{-\gamma|x|}}(\mathbb{R}^n)} \le \\ &\le C(\|\mathbf{m}_1-\mathbf{m}_2\|_{L^\infty_{e^{-\gamma|x|}}(\mathbb{R}^n)} + \|\delta\mathbf{m}_1-\delta\mathbf{m}_2\|_{L^\infty_{e^{-\gamma|x|}}(\mathbb{R}^n)} + \|v_1-v_2\|_{L^\infty_{e^{-\gamma|x|}}(\mathbb{R}^n)}) \end{aligned} \tag{5.30}$$

where the constant C depends on $\|\delta\mathbf{m}_i\|_{L^\infty(\mathbb{R}^n)}$ and $\|v_i\|_{L^\infty(\mathbb{R}^n)}$, but is independent of the concrete choice of $\mathbf{m}_i$, $\delta\mathbf{m}_i \in T_{\mathbf{m}}\mathbb{P}(L)$, v_i and L being large enough.

The proof of this result is completely analogous to Theorem 5.2 and is left to the reader.

REMARK 5.7. Obviously, the transposed operator to $\mathbb{P}_{\mathbf{m}}$ (in $L^2(\mathbb{R}^n)$) has the form

$$\mathbb{P}^*_{\mathbf{m}}v = \sum_{j=1}^{\infty}\sum_{i=1}^{k}(\phi^i_{\Gamma_j}, v)\bar\psi^j_i(\vec\Gamma). \tag{5.31}$$

Then, arguing exactly as before one can verify that the transposed projectors $\mathbb{P}^*_{\mathbf{m}}$ also satisfy the analogs of Theorems 5.2 and 5.6. In particular, analogously to (5.25), we have

$$\|\mathbb{P}^*_{\mathbf{m}}v - \sum_{j=1}^{\infty}\mathbb{P}^*_{\Gamma_j}v\|_{L^p_\theta(\mathbb{R}^n)} \le C_\varepsilon\,e^{-2(\alpha-\varepsilon)L}\,\|v\|_{L^p_\theta(\mathbb{R}^n)}$$

where $\varepsilon > 0$ and $v \in L^p_\theta(\mathbb{R}^n)$ are arbitrary and C_ε depends on ε and C_θ, but is independent of $\mathbf{m}$ and v. Moreover, the analogous estimate holds for the spaces $L^p_{b,\theta}(\mathbb{R}^n)$ as well.

We conclude this section by verifying that the tangent space $T_{\mathbf{m}}\mathbb{P}(L)$ "almost coincides" with the zero spectral subspace of the linearized operator $A + \Phi'(\mathbf{m})$. To this end, we introduce one more operator

$$\mathbb{S}(\mathbf{m})v := \mathbb{P}_{\mathbf{m}}(Av + \Phi'(\mathbf{m})v). \tag{5.32}$$

The following theorem gives the uniform smallness of the operator $\mathbb{S}(\mathbf{m})$ thus defined.

THEOREM 5.8. *Let the above assumptions hold. Then, for sufficiently large L, the operator* $\mathbb{S}(\mathbf{m})$ *depends smoothly on* $\mathbf{m}$ *and satisfies:*

$$\|\mathbb{S}(\mathbf{m})\|_{\mathcal{L}(L^\infty(\mathbb{R}^n),L^\infty(\mathbb{R}^n))} + \|\mathbb{S}'(\mathbf{m})\|_{\mathcal{L}(T_{\mathbf{m}}\mathbb{P}(L),\mathcal{L}(L^\infty(\mathbb{R}^n),L^\infty(\mathbb{R}^n)))} \le C_\varepsilon\, e^{-2(\alpha-\varepsilon)L} \tag{5.33}$$

where $\varepsilon > 0$ *is arbitrary,* C_ε *depends on* ε*, but is independent of* $\mathbf{m} \in \mathbb{P}(L)$*.*

Moreover, for sufficiently small $\gamma > 0$ *and every* $\mathbf{m}_1, \mathbf{m}_2 \in \mathbb{P}(L)$ *and* $v_1, v_2 \in L^\infty(\mathbb{R}^n)$*,*

$$\begin{aligned}\|\mathbb{S}(\mathbf{m}_1)v_1 - \mathbb{S}(\mathbf{m}_2)v_2\|_{L^\infty_{e^{-\gamma|x|}}(\mathbb{R}^n)} \le \\ \le C_\varepsilon\, e^{-\alpha L}(\|\mathbf{m}_1 - \mathbf{m}_2\|_{L^\infty_{e^{-\gamma|x|}}(\mathbb{R}^n)} + \|v_1 - v_2\|_{L^\infty_{e^{-\gamma|x|}}(\mathbb{R}^n)})\end{aligned} \tag{5.34}$$

where $\varepsilon > 0$ *is arbitrary and* C_ε *depends on* ε *and* $\|v_i\|_{L^\infty(\mathbb{R}^n)}$*,* $i = 1, 2$*, but is independent of the concrete choice of* $\mathbf{m}_i$ *and* v_i*.*

PROOF. Indeed, taking into account (5.26) and the fact that $\psi^i_{\Gamma_j}$ solves the conjugate equation, we get

$$\begin{aligned}|(\bar{\psi}^i_j(\vec{\Gamma}), Av + \Phi'(\mathbf{m})v)| = |((A^* + [\Phi'(\mathbf{m})]^*)\bar{\psi}^i_j(\vec{\Gamma}), v)| \le \\ \le |((A^* + [\Phi'(\mathbf{m})]^*)\psi^i_{\Gamma_j}, v)| + C_\varepsilon\, e^{-2(\alpha-\varepsilon)L}\,\|v\|_{L^\infty(\mathbb{R}^n)} = \\ = \|(\Phi'(\mathbf{m}) - \Phi'(V_{\Gamma_j}))\psi^i_{\Gamma_j}\|_{L^1(\mathbb{R}^n)}\|v\|_{L^\infty(\mathbb{R}^n)} + C_\varepsilon\, e^{-2(\alpha-\varepsilon)L}.\end{aligned} \tag{5.35}$$

Moreover, according to (2.32), we have

$$\begin{aligned}\|(\Phi'(\mathbf{m}) - \Phi'(V_{\Gamma_j}))\psi^i_{\Gamma_j}\|_{L^1(\mathbb{R}^n)} \le C \sup_{x\in\mathbb{R}^n}\{e^{-(\alpha-\varepsilon)|x-\xi_j|}\sum_{k\ne j} e^{-\alpha|x-\xi_k|}\} \le \\ \le C\sum_{k\ne j} e^{-(\alpha-\varepsilon)|\xi_j-\xi_k|} \le C_\varepsilon\, e^{-2(\alpha-\varepsilon)L}\end{aligned} \tag{5.36}$$

which together with (5.35) gives estimate (5.33) for $\mathbb{S}(\mathbf{m})v$. The required estimate for the derivatives and the weighted Lipschitz continuity (5.34) can be proven analogously, see the proof of Lemma 4.9 and Theorem 5.2. Theorem 5.8 is proven. □

REMARK 5.9. Analogously to the previous section, the weight $e^{-\gamma|x|}$ can be replaced in all weighted estimates of this section by any weight θ of sufficiently small exponential growth rate.

6. The multi-pulse manifold: differential equations and the cut off procedure

This section is devoted to study the differential equations on the multi-pulse manifold. We start by recalling the standard facts on the following differential equation on $\mathbb{P}(L)$:

$$\frac{d}{dt}\mathbf{m} = \mathrm{f}(t, \mathbf{m}), \quad \mathbf{m}(t_0) = \mathbf{m}_0 \tag{6.1}$$

where the unknown function $\mathbf{m}(t)$ belongs to $\mathbb{P}(L)$ for every t and the function $\mathrm{f}(t, \cdot) \in C^k(\mathbb{P}(L), L^\infty(\mathbb{R}^n))$, $k \ge 1$, such that $\mathrm{f}(t, \mathbf{m}) \subset T_{\mathbf{m}}\mathbb{P}(L)$ for every $t \in \mathbb{R}$ and every $\mathbf{m} \in \mathbb{P}(L)$. We however note that, since $\mathbb{P}(L)$ is not compact, the above assumptions are not sufficient even for the local existence of a solution. That is why we require, in addition, the global boundedness and Lipschitz continuity of the function f:

$$(6.2)\quad \| \mathrm{f}(t,\mathbf{m}_1)-\mathrm{f}(t,\mathbf{m}_2)\|_{L^\infty(\mathbb{R}^n)}\le K\|\mathbf{m}_1-\mathbf{m}_2\|_{L^\infty(\mathbb{R}^n)},\quad \|\mathrm{f}(t,\mathbf{m})\|_{L^\infty(\mathbb{R}^n)}\le C$$

where the constants K and C are independent of $\mathbf{m},\mathbf{m}_1,\mathbf{m}_2\in\mathbb{P}(L)$ and $t\in\mathbb{R}$.

THEOREM 6.1. *Let the function* f *satisfy assumptions* (6.2). *Then, for every* $\mathbf{m}_0\in\mathbb{P}(L)$ *and every* $t_0\in\mathbb{R}$, *there exists a unique local solution* $\mathbf{m}(t)$ *defined on the interval* (t_0^-,t_0^+). *Moreover, either* $t_0^-=-\infty$ *(resp.* $t_0^+=+\infty$*) or* $\mathbf{m}(t_0^-)\in\partial\mathbb{P}(L)$ *(resp.* $\mathbf{m}(t_0)\in\partial\mathbb{P}(L)$*). Moreover, this solution is Lipschitz continuous with respect to the initial data:*

$$(6.3)\qquad \|\mathbf{m}_1(t)-\mathbf{m}_2(t)\|_{L^\infty(\mathbb{R}^n)}\le e^{K|t-t_0|}\,\|\mathbf{m}_1(t_0)-\mathbf{m}_2(t_0)\|_{L^\infty(\mathbb{R}^n)}$$

for all t *such that both* $\mathbf{m}_1(t)$ *and* $\mathbf{m}_2(t)$ *are well defined.*

PROOF. The local existence and uniqueness of a solution can be verified in local coordinates exactly as in finite-dimensional case. So, we only need to verify estimate (6.3). Indeed, let $\mathbf{m}(t):=\mathbf{m}_1(t)-\mathbf{m}_2(t)$. Then, it obviously satisfies the following integral equation

$$\mathbf{m}(t)=\mathbf{m}(t_0)+\int_{t_0}^t(\mathrm{f}(s,\mathbf{m}_1(s))-\mathrm{f}(s,\mathbf{m}_2(s)))\,ds.$$

Let $t\ge t_0$ (for $t\le t_0$ the proof is analogous). Then, taking the L^∞-norm from both parts of that equality, we have

$$\|\mathbf{m}(t)\|_{L^\infty(\mathbb{R}^n)}\le\|\mathbf{m}_0(t)\|_{L^\infty(\mathbb{R}^n)}+K\int_{t_0}^t\|\mathbf{m}(s)\|_{L^\infty(\mathbb{R}^n)}\,ds$$

which, together with the Gronwall inequality, gives (6.3) and finishes the proof of the theorem. □

We recall that the above solution depends also continuously on f. Indeed, let

$$\|\mathrm{f}_1(t,\cdot)-\mathrm{f}_2(t,\cdot)\|:=\sup_{\mathbf{m}\in\mathbb{P}(L)}\|\mathrm{f}_1(t,\mathbf{m})-\mathrm{f}_2(t,\mathbf{m})\|_{L^\infty(\mathbb{R}^n)}.$$

Then, the following analog of (6.3) holds.

COROLLARY 6.2. *Let the above assumptions hold and let* $\mathbf{m}_1$ *and* $\mathbf{m}_2$ *be two solutions of* (6.1) *with the right-hand sides* f_1 *and* f_2 *respectively. Then*

$$(6.4)\quad e^{-K|t-t_0|}\,\|\mathbf{m}_1(t)-\mathbf{m}_2(t)\|_{L^\infty(\mathbb{R}^n)}\le$$
$$\le\|\mathbf{m}_1(t_0)-\mathbf{m}_2(t_0)\|_{L^\infty(\mathbb{R}^n)}+K\operatorname{sgn}(t-t_0)\int_{t_0}^t e^{-K|t_0-s|}\,\|\mathrm{f}_1(s,\cdot)-\mathrm{f}_2(s,\cdot)\|\,ds$$

for all t *such that both* $\mathbf{m}_1$ *and* $\mathbf{m}_2$ *are well-defined.*

The proof of that estimate can be obtained by Gronwall inequality exactly as in Theorem 6.1 and so omitted.

Analogously, if the Lipschitz continuity of f in weighted spaces is, a priori, known

$$(6.5)\qquad \|\mathrm{f}(t,\mathbf{m}_1)-\mathrm{f}(t,\mathbf{m}_2)\|_{L^\infty_\theta(\mathbb{R}^n)}\le K\|\mathbf{m}_1-\mathbf{m}_2\|_{L^\infty_\theta(\mathbb{R}^n)}$$

for some weight function θ of exponential growth rate, then one has the analogous Lipschitz continuity of solutions in weighted spaces.

COROLLARY 6.3. *Let θ be a weight function and* f_1 *and* f_2 *be right-hand sides of equation* (6.1) *satisfying* (6.5) *and let*

$$\|\,\mathrm{f}_1(t,\cdot)-\mathrm{f}_2(t,\cdot)\|_\theta := \sup_{\mathbf{m}\in\mathbb{P}(L)} \|\,\mathrm{f}_1(t,\mathbf{m})-\mathrm{f}_2(t,\mathbf{m})\|_{L^\infty_\theta(\mathbb{R}^n)}.$$

Then, for the associated solutions $\mathbf{m}_1$ *and* $\mathbf{m}_2$ *of equation* (6.1), *one has*

$$(6.6)\qquad \mathrm{e}^{-K|t-t_0|}\,\|\mathbf{m}_1(t)-\mathbf{m}_2(t)\|_{L^\infty_\theta(\mathbb{R}^n)} \le$$

$$\le \|\mathbf{m}_1(t_0)-\mathbf{m}_2(t_0)\|_{L^\infty_\theta(\mathbb{R}^n)} + K\,\mathrm{sgn}(t-t_0)\int_{t_0}^t \mathrm{e}^{-K|t_0-s|}\,\|\,\mathrm{f}_1(s,\cdot)-\mathrm{f}_2(s,\cdot)\|_\theta\,ds$$

for all t such that both $\mathbf{m}_1$ *and* $\mathbf{m}_2$ *are well-defined.*

Furthermore, as usual, if the function f is more regular, then the dependence of the solution $\mathbf{m}$ on the initial data is also more regular. In particular, the equation of variation associated with the solution $\mathbf{m}(t)$ of equation (6.1) reads

$$(6.7)\qquad \frac{d}{dt}\,\mathrm{w}(t) = \mathrm{f}'_{\mathbf{m}}(t,\mathbf{m}(t))\,\mathrm{w}(t), \quad \mathrm{w}(t_0)=\mathrm{w}_0\in T_{\mathbf{m}}\mathbb{P}(L)$$

or in the equivalent covariant form

$$(6.8)\qquad \mathbb{P}_{\mathbf{m}}\frac{d}{dt}\,\mathrm{w}(t) = \mathbb{P}_{\mathbf{m}}\,\mathrm{f}'_{\mathbf{m}}(t,\mathbf{m}(t))\,\mathrm{w}(t), \quad \mathrm{w}(t_0)=\mathrm{w}_0\in T_{\mathbf{m}}\mathbb{P}(L)$$

where, as usual, the expression $\nabla\,\mathrm{f}(t,\mathbf{m}) := \mathbb{P}_{\mathbf{m}}\,\mathrm{f}'_{\mathbf{m}}(t,\mathbf{m})$ can be interpreted as a covariant gradient of the vector field $\mathrm{f}(t,\cdot)$. The next standard corollary confirms that the solution w thus defined is indeed a Fréchet derivative of $\mathbf{m}(t)$ with respect to the initial data.

COROLLARY 6.4. *Let the above assumptions hold and let, in addition, the vector field* f *be Fréchet differentiable and its Fréchet derivative* $\mathrm{f}'_{\mathbf{m}}$ *is uniformly continuous on* $\mathbb{P}(L)$. *Then the solution* $\mathbf{m}(t)$ *of equation* (6.1) *is Fréchet differentiable with respect to the initial data* $\mathbf{m}_0$ *and its derivative* $\mathrm{w}(t) := D_{\mathbf{m}_0}\mathbf{m}(t)\,\mathrm{w}_0$ *solves equations* (6.7) *and* (6.8). *Analogously, if the vector field* f *is* C^k*-differentiable and its kth derivative is uniformly continuous, then the function* $\mathbf{m}_0\to\mathbf{m}(t)$ *is also* C^k*-differentiable.*

The proof of this assertion is completely analogous to the finite-dimensional case and is also left for the reader.

REMARK 6.5. It is not difficult to obtain the form of equation (6.1) in the manifold $\mathbb{B}(L)$ in terms of local coordinates Γ^i_j. Indeed, since

$$\mathbf{m}(t) := \mathbb{V}(\vec{\Gamma}(t)) = \sum_{j=1}^\infty V_{\Gamma_j(t)},$$

then

$$\frac{d}{dt}\mathbf{m}(t) = \sum_{j=1}^\infty\sum_{i=1}^k\sum_{l=1}^k \Pi_{il}(\Gamma_j)\phi^l_{\Gamma_j}\cdot\frac{d}{dt}\Gamma^i_j(t).$$

Thus, multiplying equation (6.1) by the function $\bar{\psi}^i_j(\vec{\Gamma}(t))$ (which is introduced in Lemma 5.1), integrating over $x\in\mathbb{R}^n$ and using (5.2), we have

$$(6.9)\qquad \sum_{i=1}^k \Pi_{il}\frac{d}{dt}\Gamma^i_j(t) = (\mathrm{f}(t,\mathbb{V}_{\vec{\Gamma}(t)}),\bar{\psi}^l_j(\vec{\Gamma}(t))),\quad j\in\mathbb{N},\quad l=1,\cdots,k$$

which allows us to obtain the explicit expression for (6.1) in local coordinates (we recall that the transfer matrix $\Pi_{ij}(\Gamma)$ is defined by (2.37)).

The manifold $\mathbb{P}(L)$ has a boundary and the trajectory $\mathbf{m}(t)$ of equation (6.1) can reach this boundary in finite time. As usual, this fact is very inconvenient for the center manifold constructing and some cut-off procedure is necessary. However, in contrast to the finite-dimensional case, the L^∞-distance to the boundary is not a differentiable function (even in the neighborhood of the boundary) and, thus, cannot be used for this procedure.

We overcome this difficulty by constructing, instead of a usual scalar cut-off function, a special cut-off *operator* $\operatorname{Cut}(\mathbf{m})$ acting on the vector fields on $\mathbb{P}(L)$ and define the modified version of (6.1) in the form

$$\frac{d}{dt}\mathbf{m}(t) = \operatorname{Cut}(\mathbf{m}(t))\,\mathrm{f}(t,\mathbf{m}(t)). \tag{6.10}$$

Roughly speaking, the operator $\operatorname{Cut}(\mathbf{m})$ will be identical if $\mathbf{m}$ is far from the boundary and will stop the motion of ith and jth pulse if the distance between them become close to $2L$. To be more precise, the following theorem holds.

THEOREM 6.6. *For every sufficiently large L and all $\varepsilon > 0$, there exist linear operators* $\operatorname{Cut}(\mathbf{m}) \in \mathcal{L}(L^\infty(\mathbb{R}^n), L^\infty(\mathbb{R}^n))$ *depending smoothly on* $\mathbf{m} \in \mathbb{P}(L)$ *such that,*

$$\begin{cases} 1) \ \operatorname{Cut}(\mathbf{m})\,\mathrm{w} \subset T_{\mathbf{m}}\mathbb{P}(L) \\ 2) \ \operatorname{Cut}(\mathbf{m}) = \mathbb{P}_{\mathbf{m}}, \quad \text{if } \mathbf{m} \in \mathbb{P}(L') \text{ with } L' \geq (1+\varepsilon)L \\ 3) \ \|\operatorname{Cut}(\mathbf{m})\|_{C^k(\mathbb{P}(L),\mathcal{L}(L^\infty,L^\infty))} \leq C_k. \end{cases} \tag{6.11}$$

where the constant C_k is independent of $\mathbf{m}$ and L.

Moreover, for every vector field f *satisfying the assumptions of Theorem 6.6, and every $\mathbf{m}_0 \in \mathbb{P}(L)$, the associated trajectory $\mathbf{m}(t)$ of the modified equation* (6.10) *is globally defined for all $t \in \mathbb{R}$, never reaches the boundary $\partial\mathbb{P}(L)$ (in finite time) and satisfies the analogs of estimate* (6.4).

Furthermore, the operator Cut *is Lipschitz continuous in the local topologies, i.e.*

$$\begin{aligned} &\|\operatorname{Cut}(\mathbf{m}_1)\,\mathrm{w}_1 - \operatorname{Cut}(\mathbf{m}_2)\,\mathrm{w}_2\|_{L^\infty_{e^{-\gamma|x|}}(\mathbb{R}^n)} \\ &\leq C e^{\delta L}\Big(\|\mathrm{w}_1 - \mathrm{w}_2\|_{L^\infty_{e^{-\gamma|x|}}(\mathbb{R}^n)} \\ &\qquad + (\|\mathrm{w}_1\|_{L^\infty(\mathbb{R}^n)} + \|\mathrm{w}_2\|_{L^\infty(\mathbb{R}^n)})\|\mathbf{m}_1 - \mathbf{m}_2\|_{L^\infty_{e^{-\gamma|x|}}(\mathbb{R}^n)}\Big) \end{aligned} \tag{6.12}$$

for all $\mathbf{m}_1, \mathbf{m}_2 \in \mathbb{P}(L)$ and every $\mathrm{w}_i \in T_{\mathbf{m}_i}\mathbb{P}(L)$ and sufficiently small $\gamma > 0$ and $\delta > 0$.

PROOF. We first note that, it is sufficient to construct the operator $\operatorname{Cut}(\mathbf{m})$ only on the tangent space $T_{\mathbf{m}}\mathbb{P}(L)$ (it can be then extended to all $L^\infty(\mathbb{R}^n)$ by taking the composition with the projector $\mathbb{P}_{\mathbf{m}}$.

To this end, we introduce, for every $i \in \mathbb{N}$ and some fixed $\beta > 0$, the following smooth analog of functions (3.40):

$$\widetilde{R}_i(\mathbf{m}) := \sum_{j\neq i} \theta_{\beta,\xi_i}(\xi_j) = \sum_{j\neq i} e^{-\beta\sqrt{|\xi_i-\xi_j|^2+1}}, \quad i \in \mathbb{N}, \tag{6.13}$$

where $\mathbf{m} = \mathbb{V}_{\vec{\Gamma}}$ and $\operatorname{dist}'(\xi, \Sigma)$ is defined by (4.39). Then, according to Lemma 3.12, we have

$$e^{-\beta[\operatorname{dist}'(\xi_i,\Xi)-1]} \le \widetilde{R}_i(\mathbf{m}) \le C_\beta \operatorname{dist}'(\xi_i,\Xi)^n e^{-\beta \operatorname{dist}'(\xi_i,\Xi)} . \tag{6.14}$$

Moreover, using Proposition 4.3, it is not difficult to verify that these functions are smooth with respect to $\mathbf{m}$ and the following estimate holds:

$$\|\widetilde{R}_i'(\mathbf{m})\|_{\mathcal{L}(T_{\mathbf{m}}\mathbb{P}(L),\mathbb{R})} \le C\widetilde{R}_i(\mathbf{m}) \tag{6.15}$$

where the constant C depends on β, but is independent of $\mathbf{m}$ and L and the analogous estimate hold for higher derivatives as well.

Let us now introduce the cut-off function $\widetilde{\Theta} : \mathbb{R}_+ \to \mathbb{R}_+$, $\widetilde{\Theta} \in C^\infty(\mathbb{R})$ such that

$$\widetilde{\Theta}(z) \equiv 1, \ \ z \in [0, 1/2], \quad \widetilde{\Theta}(z) \equiv 0, \ \ z \ge 1. \tag{6.16}$$

Obviously such a function exists and we have

$$C_k := \sup\{|\widetilde{\Theta}^{(k)}(z)|, \ \ z \ge 0\} < \infty. \tag{6.17}$$

For $j \in \mathbb{N}$, we introduce the following cut-off functions $\Theta_j : \mathbb{P}(L) \to \mathbb{R}, \ \ j \in \mathbb{N}$:

$$\Theta_j : \mathbb{P}(L) \to \mathbb{R}, \ \ \Theta_j(\mathbf{m}) := \widetilde{\Theta}(e^{+2\beta L}\, \tilde{R}_j(\mathbf{m})). \tag{6.18}$$

We are now ready to introduce the desired operator $\operatorname{Cut}(\mathbf{m})$. To this end, we rewrite equation (6.1) in the equivalent form (6.9) and defined the desired modified equation as follows:

$$\sum_{i=1}^{k} \Pi_{il}(\Gamma_j)\frac{d}{dt}\Gamma_j^i = \Theta_j(\mathbf{m})(\mathrm{f}(t,\mathbf{m}), \bar{\psi}_j^l(\mathbf{m})). \tag{6.19}$$

Returning now to the $\mathbf{m}$ variable, we have

$$\frac{d}{dt}\mathbf{m} = \sum_{j\in\mathbb{N}}\sum_{l=1}^{k} \Theta_j(\mathbf{m})(\mathrm{f}(t,\mathbf{m}), \bar{\psi}_j^l(\mathbf{m}))\phi_{\Gamma_j}^l \tag{6.20}$$

and, consequently,

$$\operatorname{Cut}(\mathbf{m})\,\mathrm{w} := \sum_{j\in\mathbb{N}}\sum_{i=1}^{k} \Theta_j(\mathbf{m})(\mathrm{w}, \bar{\psi}_j^i(\mathbf{m}))\phi_{\Gamma_j}^i . \tag{6.21}$$

We claim that the operator $\operatorname{Cut}(\mathbf{m})$, thus defined, satisfies all of the assertions of the theorem. Indeed, the first assertion of (6.11) is obvious. Let us verify the second one. Let $L' = L'(L) > L$ solves

$$C_\beta (L')^n e^{-2\beta L'} = 1/2\, e^{-2\beta L} .$$

Then, on the one hand, for sufficiently large $L > L_0(\varepsilon)$, $L' \le L' := (1+\varepsilon)L$. On the other hand, due to (6.15) and (6.16), we have

$$\Theta_j(\mathbf{m}) \equiv 1, \ \ \forall \mathbf{m} \in \mathbb{P}(L') \subset \mathbb{P}(L), \ \ j \in \mathbb{N}. \tag{6.22}$$

It remains to note that (6.21) and (6.22) together with the orthogonality relation (5.2) imply that

$$\operatorname{Cut}(\mathbf{m})\,\mathrm{w} = \mathrm{w}, \quad \forall \mathrm{w} \in T_{\mathbf{m}}\mathbb{P}(L').$$

Thus, the second assertion of (6.11) is verified. Furthermore, the smoothness of the operator $\operatorname{Cut}(\mathbf{m})$ follows from the explicit formula (6.21) and estimate (6.11)(3) follows from (6.15) and (6.17) and, therefore, assertions (6.11) are verified.

Let us now verify that the trajectory $\mathbf{m}(t)$ of the cutted-off equation (6.10) cannot reach the boundary $\partial\mathbb{P}(L)$ in a finite time. Indeed, let the distance between two pulses $V_{\Gamma_{j_1}(T)}$ and $V_{\Gamma_{j_2}(T)}$ is less than $2L+1$ at some time T. Then, according to (6.16) and left inequality of (5.6)

$$\Theta_{j_1}(\mathbf{m}(T)) = \Theta_{j_2}(\mathbf{m}(T)) = 0$$

and, consequently, these pulses do not move and the distance between them cannot decrease under the time evolution. Thus, the trajectory $\mathbf{m}(t)$ cannot indeed reach the boundary in a finite time and, therefore, due to Theorem 6.1, is defined for all $t \in \mathbb{R}$.

Thus, it only remains to verify the weighted estimate (6.12). This estimate can be easily verified arguing as in the proof of Theorem 5.2 and so is left to the reader (the only difference here is that we do not have the analog of (6.15) in weighted norms and, as a result, the weighted Lipschitz constant in (6.12) contains exponentially growing term $e^{\delta L}$ where the constant $\delta = \delta(\gamma)$ can chosen arbitrarily small if γ is small enough). Theorem 6.6 is proven. □

7. Slow evolution of multi-pulse profiles: linear case

In this section, we begin the study the dynamics, generated by equation (2.1) in a neighborhood of the multi-pulse manifold $\mathbb{P}(L)$. We start with the following result which gives the nonlinear "orthogonal" projectors to the manifold $\mathbb{P}(L)$.

THEOREM 7.1. *Let the assumptions of Section 2 hold and let L be large enough. Then, there exists $\delta > 0$ and the nonlinear smooth function $\pi : \mathcal{O}_\delta(\mathbb{P}(L+1)) \to \mathbb{P}(L)$ (where the δ-neighborhood is taken in the L^∞-topology) such that*

$$\mathbb{P}_{\pi(v)}(v - \pi(v)) \equiv 0, \quad \text{for all } v \in \mathcal{O}_\delta(\mathbb{P}(L+1)) \tag{7.1}$$

The function $\pi(v)$ is uniquely defined by this equation,

$$\pi(\mathbf{m}) = \mathbf{m}, \quad \pi'(\mathbf{m}) = \mathbb{P}_{\mathbf{m}}, \quad \forall \mathbf{m} \in \mathbb{P}(L+1) \tag{7.2}$$

and moreover, its C^k-derivatives are uniformly bounded for any $k \in \mathbb{N}$:

$$\|\pi\|_{C^k(\mathcal{O}_\delta(\mathbb{P}(L)),\mathbb{P}(L))} \le C_k. \tag{7.3}$$

Finally, the constants δ and C_k are independent of L.

PROOF. As usual, the existence of such projector can be obtained by the implicit function theorem. In order to show that, we first note that the required projector can be constructed locally, for the δ-neighborhood of every point $\mathbf{m}_0 \in \mathbb{P}(L+1)$. Indeed, the local uniqueness of it will be guaranteed by the implicit function theorem and the global uniqueness follows immediately from the local one.

Thus, it is sufficient to verify the existence of π only in the neighborhood $\mathcal{O}_\delta(\mathbf{m}_0)$ for some $\mathbf{m}_0 \in \mathbb{P}(L+1)$. Then, $v = \mathbf{m}_0 + w$ with $\|w\|_{L^\infty} \le \delta$ and the required $\mathbf{m} := \pi(v)$ should be found from the equation

$$\mathbb{P}_{\mathbf{m}}(\mathbf{m}_0 + w - \mathbf{m}) = 0.$$

Or, rewriting it in the local coordinates Γ_j^i near $\vec{\Gamma}^0 = [\mathbb{V}]^{-1}\mathbf{m}_0$ (see Remark 4.5), we will have

$$(\mathbb{V}_{\vec{\Gamma}^0} + w - \mathbb{V}_{\vec{\Gamma}}, \vec{\psi}_j^i(\vec{\Gamma})) = 0, \quad j \in \mathbb{N}, \quad i = 1, \cdots, k.$$

The last equation can be easily solved by the implicit function theorem in the space l^∞. Indeed, let $F : l^\infty \times L^\infty(\mathbb{R}^n) \to \mathbb{R}^\infty$ be defined via

$$F(\vec{\Gamma}, w)^i_j := (\mathbb{V}_{\vec{\Gamma}^0} + w - \mathbb{V}_{\vec{\Gamma}}, \bar{\psi}^i_j(\vec{\Gamma})) \tag{7.4}$$

(where, for simplicity, we have identified the point $\vec{\Gamma} \in \mathbb{B}(L)$ with its local coordinates near $\vec{\Gamma}^0$).

Obviously, the function F depends smoothly on $\vec{\Gamma}$ and w. Moreover, $F(\vec{\Gamma}_0, 0) = 0$ and

$$(F'_{\vec{\Gamma}}(\vec{\Gamma}^0, 0)\delta\Gamma)^i_j = (-\sum_{l,r,m} \Pi_{lm}(\Gamma^0_r)\phi^l_{\Gamma^0_r}\delta\Gamma^m_r, \bar{\psi}^i_j(\vec{\Gamma}_0)) = -\delta\Gamma^i_j$$

where we have implicitly used the orthogonality relations (5.2) and the fact that the local coordinates on G_0 near $\Gamma = \Gamma^0$ can be chosen in such way that $\Pi(\Gamma^0) = \mathrm{Id}$, see Remark 2.3. Thus, $D_{\vec{\Gamma}}F(\vec{\Gamma}^0, 0) = -\mathrm{Id}$ and, consequently, due to the implicit function theorem, equation $F(\vec{\Gamma}, w) = 0$ has a unique solution $\vec{\Gamma} = \vec{\Gamma}(w)$ if $\|w\|_{L^\infty} \le \delta$ is small enough. Moreover,it is not difficult to see that the constant δ is independent also on $\vec{\Gamma}^0 \in \mathbb{B}(L+1)$. Thus, the required projector $\pi : \mathcal{O}_\delta(\mathbb{P}(L+1)) \to \mathbb{P}(L)$ is constructed (we note that $\vec{\Gamma}(w)$ does not necessarily belong to $\mathbb{P}(L+1)$, but it always belong to $\mathbb{P}(L)$ if $\delta > 0$ is small enough). Formulae (7.2) follow from the basic equation (7.1) and the differentiability and estimates (7.3) are simple corollaries of the implicit function theorem. Theorem 7.1 is proven. □

REMARK 7.2. The above theorem shows, in particular, that every point $u \in \mathcal{O}_\delta(\mathbb{P}(L+\varepsilon))$, $\varepsilon > 0$, can be uniquely decomposed in a sum

$$u = v + \mathbf{m}, \quad \mathbf{m} \in \mathbb{P}(L), \quad \mathbb{P}_{\mathbf{m}} v = 0$$

where $\mathbf{m} := \pi(u)$. In turns, this splitting shows that $\mathbb{P}(L)$ is a submanifold in $L^\infty(\mathbb{R}^n)$ where the neighborhood of $\mathbf{m} \in \mathbb{P}(L)$ locally looks like $\ker P_{\mathbf{m}} \times T_{\mathbf{m}}\mathbb{P}(L) \sim \ker P_{\mathbf{m}} \times l^\infty$.

According to Theorem (7.1), it seems natural to seek for the solution $u(t)$ of problem (2.1) in the form

$$u(t) = v(t) + \mathbf{m}(t), \quad \mathbf{m}(t) \in \mathbb{P}(L), \quad \mathbb{P}_{\mathbf{m}(t)} v(t) \equiv 0 \tag{7.5}$$

where $\mathbf{m}(t)$ solves the appropriate equation on $\mathbb{P}(L)$ and $v(t)$ is small and satisfies some equation on the "normal" bundle of $\mathbb{P}(L)$ generated by $\ker P_{\mathbf{m}}$. Differential equations on $\mathbb{P}(L)$ have been studied in the previous section and the aim of this section is to study the linearized equations for $v(t)$ (the nonlinear case will be considered in the next section). To be more precise, we will study the following linear problem

$$\partial_t v + A_0 v + \Phi'(\mathbf{m}(t))v = h(t), \tag{7.6}$$

where $\mathbf{m}(t)$ is an arbitrary "slow" trajectory on $\mathbb{P}(L)$ and find necessary and sufficient conditions on $h(t)$ which allow to satisfy the additional condition $\mathbb{P}_{\mathbf{m}(t)} v(t) \equiv 0$. These conditions will be used in the next section in order to deduce the equations for the pulse evolution.

Analogously to (3.18), it is convenient to eliminate the "neutral" modes of equation (7.6) by adding the artificial projector as follows:

$$\partial_t v + A_0 v + \Phi'(\mathbf{m}(t))v + \mathbb{P}_{\mathbf{m}(t)} v = h(t), \tag{7.7}$$

We start our consideration by the case of standing pulses $\mathbf{m}(t) \equiv \mathbf{m} \in \mathbb{P}(L)$

$$\partial_t v + A_0 v + \Phi'(\mathbf{m})v + \mathbb{P}_{\mathbf{m}} v = h(t). \tag{7.8}$$

and then treat the case of slowly moving pulses via a perturbation technique. The next proposition gives the solvability (7.8) if L is large enough.

PROPOSITION 7.3. *Let the operators A_0 and Φ satisfy all of the assumptions of Section 2. Then, for every $1 < p < \infty$, there exists a (large) constant L_0 and a (small) positive constant ε_0 such that, for every $L \geq L_0$, every $\mathbf{m} \in \mathbb{P}(L)$ and every weight function θ with exponential growth rate $\varepsilon \leq \varepsilon_0$, equation (7.8) is uniquely solvable for every $h \in L^p_\theta(\mathbb{R}^{n+1})$ in the class $W^{(1,2l),p}_\theta(\mathbb{R}^{n+1})$ and the following estimate holds:*

$$\|v\|_{W^{(1,2l),p}_\theta(\mathbb{R}^{n+1})} \leq C\|h\|_{L^p_\theta(\mathbb{R}^{n+1})} \tag{7.9}$$

where the constant C depends only on C_θ and is independent of L, ε, $\mathbf{m} \in (L)$ and of the concrete choice of the weight θ. Moreover, the analogous result holds for the adjoint equation:

$$-\partial_t w + A_0^* w + [\Phi'(\mathbf{m})]^* w + \mathbb{P}^*_{\mathbf{m}} w = h \tag{7.10}$$

and for the weighted spaces $L^p_{b,\theta}$.

PROOF. We first note that it is sufficient to verify the assertion of Proposition 7.3 for the non-weighted case $\theta = 1$ only. The case of general weights can be deduced from this particular case using (5.23), (5.24) exactly as in Proposition 3.10. Thus, we assume from now on that $h \in L^p(\mathbb{R}^{n+1})$ and $\theta = 1$.

We first construct an approximate solution of equation (7.8). We seek for that approximate solution in the form

$$\tilde{v}(t,x) := v_0(t,x) + \sum_{i=1}^{\infty} v_i(t,x) \tag{7.11}$$

where v_i solves

$$\partial_t v_i + A_0 v_i + (F_{\Gamma_i} + \mathbb{P}_{\Gamma_i})v_i = h_i(t,x) := h(t,x)\chi_{B^L_{\xi_i}}(x), \tag{7.12}$$

$\chi_V(x)$ is the standard characteristic function of the set $V \subset \mathbb{R}^n$ and the remainder v_0 satisfies

$$\partial_t v_0 + A_0 v_0 = h_0(t,x) := h(t,x) - \sum_{i=1}^{\infty} h_i(t,x). \tag{7.13}$$

Let us consider now the sequence of sets $V_i := B^L_{\xi_i}$, $i = 1, \cdots, \infty$ and $V_0 := \mathbb{R}^n \backslash (\sum_{i=1}^{\infty} V_i)$ and the associated special weights

$$\theta_{i,\beta}(x) := e^{\beta \operatorname{dist}(x,V_i)}, \quad i = 0, \cdots, \infty. \tag{7.14}$$

Then, obviously, the weights $\theta_{i,\beta}(x)$ are exponential with growth rate β and, since the sequence Γ_j is L-separated, we have $V_i \cap V_j = \varnothing$ for $i \neq j$. Moreover, according to Proposition 3.10, for sufficiently small β and ε, we have

$$\|v_i\|_{W^{(1,2l),p}_{\theta_{i,\beta}(x)\,e^{-\varepsilon|z-z_0|}}(\mathbb{R}^{n+1})} \leq C\|h\|_{L^p_{e^{-\varepsilon|z-z_0|}}(\mathbb{R}\times V_i)}, \quad i = 0, \cdots, \infty \tag{7.15}$$

where the constant C is independent of L, Γ_i and $z_0 := (t_0, x_0) \in \mathbb{R}^n$. We have used here that $\theta_{i,\beta}(x) \equiv 1$ if $x \in V_i$. Let us now consider the approximation error operator

$$\mathcal{R}h := \partial_t \tilde{v} + A_0 \tilde{v} + \Phi'(\mathbf{m})\tilde{v} + \mathbb{P}_{\mathbf{m}}\tilde{v} - h. \tag{7.16}$$

Then, since $h = \sum_{i=0}^{\infty} h_i$, we have

$$\mathcal{R}h = [\Phi'(\mathbf{m}) - \sum_{j=1}^{\infty} \Phi'(V_{\Gamma_j})]\tilde{v} + [\mathbb{P}_{\mathbf{m}} - \sum_{j=1}^{\infty} \mathbb{P}_{\Gamma_j}]\tilde{v} + \sum_{i=0}^{\infty} \mathcal{R}_i v_i \tag{7.17}$$

where

$$\mathcal{R}_i u := \sum_{j=1, j\neq i}^{\infty} (F_{\Gamma_j} + \mathbb{P}_{\Gamma_j})u, \quad F_{\Gamma_j} := \Phi'(V_{\Gamma_j}).$$

The following lemma shows that the error operator $\mathcal{R}$ is uniformly small if L is large enough. The proof of Proposition 7.3 will be continued after this.

LEMMA 7.4. *Let the above assumptions hold. Then, for sufficiently large L,*

$$\|\mathcal{R}h\|_{L^p(\mathbb{R}^{n+1})} \le C\,\mathrm{e}^{-\delta L}\, \|h\|_{L^p(\mathbb{R}^{n+1})} \tag{7.18}$$

where the positive constants C and δ are independent of L, h and Γ_j.

PROOF. We first note that, according to estimates (7.15) and Proposition 3.13, we have

$$\|\tilde{v}\|_{W^{(1,2l),p}(\mathbb{R}^{n+1})} \le C\|h\|_{L^p(\mathbb{R}^{n+1})} \tag{7.19}$$

where the constant C is independent of $L \ge L_0$, h and $\mathbf{m} \in \mathbb{P}(L)$. Then, arguing analogously to Lemma 4.8, we deduce that

$$\|[\Phi'(\mathbf{m}) - \sum_{j=1}^{\infty} \Phi'(V_{\Gamma_j})]\tilde{v}\|_{L^p(\mathbb{R}^{n+1})} \le \tag{7.20}$$
$$\le C\,\mathrm{e}^{-\alpha L}\, \|\tilde{v}\|_{W^{(1,2l),p}(\mathbb{R}^{n+1})} \le C'\,\mathrm{e}^{-\alpha L}\, \|h\|_{L^p(\mathbb{R}^{n+1})}.$$

Moreover, due to estimate (5.25), we also have

$$\|[\mathbb{P}_{\mathbf{m}} - \sum_{j=1}^{\infty} \mathbb{P}_{\Gamma_j}]\tilde{v}\|_{L^p(\mathbb{R}^{n+1})} \le C\,\mathrm{e}^{-\alpha L}\, \|h\|_{L^p(\mathbb{R}^{n+1})}. \tag{7.21}$$

Thus, the first two terms in (7.17) satisfy (7.18) and we only need to estimate the sum $\sum_{i=0}^{\infty} \mathcal{R}_i v_i$.

We are going to do so using Proposition 3.13. Indeed, according to the first estimate of (3.33) and estimates (3.6) and (3.54) and the fact that the weights $\theta = \theta_{i,\beta}$ are exponential with growth rate β and the constant C_θ is uniformly bounded with respect to i, we have that, for sufficiently small positive β and ε:

$$\|(F_{\Gamma_j} + \mathbb{P}_{\Gamma_j})v_i\|_{L^p_{\theta_{i,\beta}(x)\,\mathrm{e}^{\alpha|x-\xi_j|/2-\varepsilon|x-x_0|}}(\mathbb{R}^n)} \le C\|v_i\|_{W^{2l-1,p}_{\theta_{i,\beta}(x)\,\mathrm{e}^{-\varepsilon|x-x_0|}}(\mathbb{R}^n)} \tag{7.22}$$

where the constant C is independent of i, j, Γ_j and $x_0 \in \mathbb{R}^n$. We now note that assumption (3.1) applied to weight functions $\theta_{i,\beta}(x)$ imply that

$$\theta_{i,\beta}(x) \ge C_\theta^{-1}\theta_{i,\beta}(\xi_j)\,\mathrm{e}^{-\beta|x-\xi_j|}. \tag{7.23}$$

Moreover, it follows from the explicit form of the weights $\theta_{i,\beta}(x)$ and from the fact that Γ_j are $2L$-separated that

$$\theta_{i,\beta}(\xi_j) \geq e^{\beta L}, \quad \forall i \neq j \tag{7.24}$$

and, consequently, if $\beta < \alpha/4$, we have for $i \neq j$

$$\theta_{i,\beta}(x)\, e^{\alpha|x-\xi_j|/2} \geq C\, e^{\beta L}\, e^{\alpha|x-\xi_j|/4}$$

and, thus, estimate (7.22) implies that

$$\|(F_{\Gamma_j} + \mathbb{P}_{\Gamma_j})v_i\|_{L^p_{e^{\alpha|x-\xi_j|/4-\varepsilon|x-x_0|}}(\mathbb{R}^n)} \leq C'\, e^{-\beta L}\, \|v_i\|_{W^{2l-1,p}_{\theta_{i,\beta}(x)\, e^{-\varepsilon|x-x_0|}}(\mathbb{R}^n)} \tag{7.25}$$

for all $j \neq i$ (and with the constant C' independent of L, i, j and Γ_j. Estimate (7.25) allows to apply Proposition 3.13 for estimating $\mathcal{R}_i$ which gives

$$\|\mathcal{R}_i v_i\|_{L^p_\theta(\mathbb{R}^n)} \leq C''\, e^{-\beta L}\, \|v_i\|_{W^{2l-1,p}_{\theta_{i,\beta}\theta}(\mathbb{R}^n)} \tag{7.26}$$

which is valid for every weight function of a sufficiently small exponential growth rate $\varepsilon \leq \varepsilon_0$. In particular, for $i = 0$, (7.26) together with (7.15) and integration over $t \in \mathbb{R}$ imply that

$$\|\mathcal{R}_0 v_0\|_{L^p(\mathbb{R}^{n+1})} \leq C_1\, e^{-\beta L}\, \|h\|_{L^p(\mathbb{R}^{n+1})} \tag{7.27}$$

and, thus, we it remains to estimate an infinite sum $\sum_{i=1}^{\infty} \mathcal{R}_i v_i$ using again Proposition 3.13. To this end, we take $\theta = \theta_{i,\beta_0}(x)\, e^{-\varepsilon|x-x_0|}$ in estimate (7.26) with positive β_0 and ε satisfying $\beta_0 + \varepsilon \leq \varepsilon_0$. Then, using the fact that $\theta_{i,\beta_0}(x) \geq e^{-\beta_0 L}\, e^{\beta_0|x-\xi_i|}$, we obtain from (7.26) that

$$\|\mathcal{R}_i v_i\|^p_{L^p_{e^{\beta_0|x-\xi_i|-\varepsilon|x-x_0|}}(\mathbb{R}^n)} \leq (C_2\, e^{-(\beta-\beta_0)L})^p \|v_i\|^p_{W^{2l-1,p}_{\theta_{i,\beta+\beta_0}(x)\, e^{-\varepsilon|x-x_0|}}(\mathbb{R}^n)} \tag{7.28}$$

where the constant C_2 is independent of L, $z_0 := (t_0, x_0) \in \mathbb{R}^{n+1}$ and Γ_j. Multiplying (7.28) by $e^{-p\varepsilon|t-t_0|}$, integrating over $t \in \mathbb{R}$ and using (7.15) (where $\theta_{i,\beta}$ is now replaced by $\theta_{i,\beta+\beta_0}$, we finally deduce

$$\|\mathcal{R}_i v_i\|_{L^p_{e^{\beta_0|x-\xi_j|-\varepsilon|z-z_0|}}(\mathbb{R}^{n+1})} \leq C\, e^{-(\beta-\beta_0)L}\, \|h\|_{L^p_{e^{-\varepsilon|z-z_0|}}(\mathbb{R}^{n+1})}. \tag{7.29}$$

Estimate (7.29) allow to apply Proposition 3.13 (see also Remark 3.15) for estimating the sum $\sum_{i=1}^{\infty} \mathcal{R}_i v_i$ which together with (7.27) imply (7.18) and finishes the proof of the lemma. □

We are now ready to finish the proof of Proposition 7.3. To this end, we note that, due to estimate (7.19), the linear operator $\mathbb{T} : h \to \tilde{v}$ is well-defined and uniformly bounded. Fix now the constant $L_0 > 0$ large enough that the $(L^p \to L^p)$-norm of the error operator $\mathcal{R}$ will be less than $1/2$. Then, the exact solution of problem (7.8) can be found by the following Neumann series:

$$v := \mathbb{T} \circ (\sum_{i=0}^{\infty} \mathcal{R}^i)v. \tag{7.30}$$

The convergence of the series is guaranteed by the fact that $\mathcal{R}$ is a contraction and estimate (7.19) guarantees that the solution thus obtained satisfies estimate (7.9) with $\theta = 1$. Thus, the existence of the required solution v is verified. In order to verify its uniqueness, it is sufficient to prove that the adjoint equation (3.3) is also solvable (in the same class) for every $h \in L^p(\mathbb{R}^{n+1})$, but this fact can be verified exactly in the same way as for equation (3.1). Proposition 7.3 is proven. □

We are now ready to study the case of slowly moving pulses. To this end, we assume that we are given a trajectory $\mathbf{m}(t) \in \mathbb{P}(L)$ for all $t \in \mathbb{R}$ and depending *slowly* on time, i.e. there exists a small positive constant ν (which will be specified below) such that

$$\|\frac{d}{dt}\mathbf{m}(t)\|_{L^\infty(\mathbb{R}^n)} \le \nu, \quad t \in \mathbb{R}. \tag{7.31}$$

We now consider the nonautonomous equation (7.7) where the pulse curve $\mathbf{m}(t)$ satisfies (7.31). Since the Sobolev's norms are equivalent on the manifold $\mathbb{P}(L)$, see Corollary 4.4, then (7.31) implies

$$\|\mathbf{m}'(t)\|_{C^k(\mathbb{R}^n)} \le C_k\|\mathbf{m}'(t)\|_{L^\infty(\mathbb{R}^n)} \le C_k\nu \tag{7.32}$$

for all $k \in \mathbb{N}$.

The following theorem, which gives the unique solvability of equation (7.7) is considered as the main result of this section.

THEOREM 7.5. *Let the operators A_0 and Φ satisfy all of the assumptions formulated in Section 2. Then, there exist positive ν_0, L_0 and ε_0 such that for any $L \ge L_0$, for any curve $\mathbf{m} : \mathbb{R} \to \mathbb{P}(L)$ satisfying (7.31) with $\nu \le \nu_0$ and every weight function θ of exponential growth rate $\varepsilon \le \varepsilon_0$ equation (7.7) is uniquely solvable in $W_\theta^{(1,2l),p}(\mathbb{R}^{n+1})$ for every $h \in L_\theta^p(\mathbb{R}^{n+1})$ and the following estimate holds:*

$$\|v\|_{W_\theta^{(1,2l),p}(\mathbb{R}^{n+1})} \le C\|h\|_{L_\theta^p(\mathbb{R}^{n+1})} \tag{7.33}$$

where the constant C depends on C_θ, but is independent of h, L, ν and the concrete choice of the curve $\mathbf{m}(t)$. Moreover, the analogous estimates hold also for the adjoint equation of (7.7) and for the spaces $L_{b,\theta}^p$.

PROOF. First of all, we note that (as in Proposition 7.3), it is sufficient to verify (7.33) for the non-weighted case $\theta = 1$ only. Estimate (7.33) for general weights can be deduced from the non-weighted one using estimate (5.24) exactly as in Proposition 7.3 and Proposition 3.5. Thus, we assume below that $h \in L^p(\mathbb{R}^{n+1})$.

As in the proof of Proposition 7.3, we are going to construct the approximate solution $\tilde v(t,x)$ of problem (7.7). To this end, for every $\mathbf{m} \in \mathbb{P}(L)$ we introduce the solution operator $\mathbb{L}_\mathbf{m} : L^p(\mathbb{R}^{n+1}) \to W^{(1,2l),p}(\mathbb{R}^{n+1})$ via

$$\mathbb{L}_\mathbf{m} h := \bar v \tag{7.34}$$

where $\bar v$ is a unique solution of equation (3.1) (with standing pulses) constructed in Proposition 7.3. For every $s \in \mathbb{R}$, we define now

$$w(s,t,x) := (\mathbb{L}_{\mathbf{m}(s)}h)(t,x) \tag{7.35}$$

(we emphasize that, for every $s \in \mathbb{R}$, we solve equation (7.8) with *standing* pulses $\mathbf{m}(s)$) and then define the approximative solution $\tilde v(t,x)$ of (7.7) by the following expression:

$$\tilde v(t,x) := w(t,t,x). \tag{7.36}$$

We now need to deduce some estimates for the functions (7.35) and (7.36). To this end, we recall that $\mathbf{m}(s)$ is assumed to belong to $\mathbb{P}(L)$ for every $s \in \mathbb{R}$ and, consequently, due to Proposition 7.3,

$$\|w(s,\cdot,\cdot)\|_{W_\theta^{(1,2l),p}(\mathbb{R}^{n+1})} \le C\|h\|_{L_\theta^p(\mathbb{R}^{n+1})} \tag{7.37}$$

where the constant C is independent of s and $\vec{\Gamma}$ and θ is a weight function with a sufficiently small exponential growth rate. We are now going to estimate the function $\tilde{w}(s,t,x) := \partial_s w(s,t,x)$ which obviously satisfies the following equation:

$$\partial_t\tilde{w}+A_0\tilde{w}+\Phi'(\mathbf{m}(s))\tilde{w}+\mathbb{P}_{\mathbf{m}(s)}\tilde{w} = -\Phi''(\mathbf{m}(s))\mathbf{m}'(s)w - D_{\mathbf{m}}\mathbb{P}_{\mathbf{m}(s)}\mathbf{m}'(s)w. \tag{7.38}$$

Using now estimates (5.2), (7.32) and (7.37), we deduce

$$\|\Phi''(\mathbf{m}(s))\mathbf{m}'(s)w + \mathbb{P}'_{\mathbf{m}(s)}\mathbf{m}'(s)w\|_{L^p_\theta(\mathbb{R}^{n+1})} \le \\ \le C\nu\|w\|_{W^{(1,2l),p}_\theta(\mathbb{R}^{n+1})} \le C'\nu\|h\|_{L^p_\theta(\mathbb{R}^{n+1})}. \tag{7.39}$$

Applying now estimate (7.9) to equation (7.38) and using (7.39), we infer

$$\|\tilde{w}(s,\cdot,\cdot)\|_{W^{(1,2l),p}_\theta(\mathbb{R}^{n+1})} \le C\nu\|h\|_{L^p_\theta(\mathbb{R}^{n+1})} \tag{7.40}$$

where the constant C is independent of ν, $\mathbf{m}$, h and the concrete form of the weight θ. We are now going to verify that

$$\|\tilde{v}\|_{W^{(1,2l),p}(\mathbb{R}^{n+1})} \le C\|h\|_{L^p(\mathbb{R}^{n+1})}. \tag{7.41}$$

To this end, we observe that (7.37) implies the following estimate:

$$\int_m^{m+1} \|\partial_t w(s,r,\cdot)\|^p_{L^p(\mathbb{R}^n)} + \|w(s,r,\cdot)\|^p_{W^{2l,p}(\mathbb{R}^n)}\,dr \le \\ \le C\int_{\mathbb{R}} e^{-p\varepsilon|m-t|}\,\|h(t)\|^p_{L^p(\mathbb{R}^n)}\,dt \tag{7.42}$$

for sufficiently small $\varepsilon > 0$ and the constant C independent of $m \in \mathbb{R}$. Analogously, (7.40) gives

$$\|\tilde{w}(s,m,\cdot)\|^p_{L^p(\mathbb{R}^n)} + \int_m^{m+1} \|\partial_t\tilde{w}(s,r,\cdot)\|^p_{L^p(\mathbb{R}^n)} + \|\tilde{w}(s,r,\cdot)\|^p_{W^{2l,p}(\mathbb{R}^n)}\,dr \le \\ \le (C\nu)^p\int_{\mathbb{R}} e^{-p\varepsilon|m-t|}\,\|h(t)\|^p_{L^p(\mathbb{R}^n)}\,dt. \tag{7.43}$$

Combining estimates (7.42) and (7.43) and using the obvious formula

$$|w(t,t,x) - w(m,t,x)| \le \int_m^{m+1} |\tilde{w}(s,t,x)|\,ds, \quad t \in [m, m+1]$$

we obtain

$$\int_m^{m+1} \|\partial_t\tilde{v}(s,\cdot)\|^p_{L^p(\mathbb{R}^n)} + \|\tilde{v}(s,\cdot)\|^p_{W^{2l,p}(\mathbb{R}^n)}\,dr \le C\int_{\mathbb{R}} e^{-p\varepsilon|m-t|}\,\|h(t)\|^p_{L^p(\mathbb{R}^n)}\,dt. \tag{7.44}$$

Summing inequalities (7.44) for all $m \in \mathbb{Z}$, we finally deduce (7.41).

Now it is not difficult to finish the proof of the theorem. To this end we consider, as in the proof of Proposition 7.3, the approximation error operator $\mathcal{R}$: $L^p(\mathbb{R}^{n+1}) \to L^p(\mathbb{R}^{n+1})$ defined via

$$(\mathcal{R}h)(t,x) := \partial_t\tilde{v} + A_0\tilde{v} + \Phi'(\mathbf{m}(t))\tilde{v} + P_{\mathbf{m}(t)}\tilde{v} - h = \tilde{w}(t,t,x) \tag{7.45}$$

where we used that

$$\partial_t\tilde{v} = [\partial_s w(s,t,\cdot) + \partial_t w(s,t,\cdot)]_{s=t} = \partial_t w(t,t,\cdot) + \tilde{w}(t,t,\cdot).$$

It now follows from (7.43) that

$$\|\mathcal{R}h\|_{L^p(\mathbb{R}^{n+1})} \le C\nu\|h\|_{L^p(\mathbb{R}^{n+1})}, \tag{7.46}$$

where the constant C is independent of ν and of the concrete choice of $\Gamma_j(t)$. Thus, $\mathcal{R}$ is a contraction for sufficiently small ν_0 (e.g., if $C\nu_0 \le 1/2$) and, consequently, the exact solution $v(t,x)$ of (7.7) can be found by the Neumann series (7.30) where the operator $\mathbb{T}$ maps h to $\tilde{v}$. Moreover, due to estimate (7.41) the solution v thus constructed satisfies (7.33) with $\theta = 1$. In order to verify the uniqueness of that solution, it is sufficient to verify that the adjoint equation

$$(7.47) \qquad -\partial_t w + A_0^* w + [\Phi'(\mathbf{m}(t))]^* w + \mathbb{P}^*_{\mathbf{m}(t)} w = h.$$

also possesses a solution $w \in W^{(1,2l),p}(\mathbb{R}^{n+1})$ for every $h \in L^p(\mathbb{R}^{n+1})$ which can be proved exactly as for equation (7.7). Theorem 7.5 is proven. □

Thus, due to Theorem 7.5, the solution operator $\mathbb{T}_{\mathbf{m}}$ of problem (7.7) is well-defined as a linear operator from $L^p_\theta(\mathbb{R}^{n+1})$ to $W^{(1,2l),p}_\theta(\mathbb{R}^{n+1})$ and from $L^p_{b,\theta}(\mathbb{R}^{n+1})$ to $W^{(1,2l),p}_{b,\theta}(\mathbb{R}^{n+1})$ for all $1 < p < \infty$ and every weight function θ of sufficiently small exponential growth rate by the following expression:

$$(7.48) \qquad \mathbb{T}_{\mathbf{m}} h := v$$

where v solves (7.7) with the right-hand side h. The next lemma shows that this operator depends continuously on the curve $\mathbf{m}(t)$

LEMMA 7.6. *Let the assumptions of Theorem 7.5 hold and let $h \in L^p_{\mathrm{b}}(\mathbb{R}^{n+1})$. Then, for every weight function θ of sufficiently small exponential growth and every two curves $\mathbf{m}_1(t)$ and $\mathbf{m}_2(t)$ satisfying the assumptions of Theorem 7.5, the following estimate holds:*

$$(7.49) \qquad \|\mathbb{T}_{\mathbf{m}_1} h - \mathbb{T}_{\mathbf{m}_2} h\|_{W^{(1,2l),p}_{b,\theta}(\mathbb{R}^{n+1})} \le C\|h\|_{L^p_{\mathrm{b}}(\mathbb{R}^n)} \|\mathbf{m}_1 - \mathbf{m}_2\|_{L^\infty_\theta(\mathbb{R}^{n+1})}$$

where the constant C is independent of h and the functions $\mathbf{m}_1(\cdot)$ and $\mathbf{m}_2(\cdot)$ satisfying the assumptions of Theorem 7.5.

PROOF. Indeed, let $w_i := \mathbb{T}_{\mathbf{m}_i} h$. Then the difference $w := w_1 - w_2$ satisfies the equation:

$$(7.50) \quad \partial_t w + A_0 w + \Phi'(\mathbf{m}_1(t))w + \mathbb{P}_{\mathbf{m}_1(t)} w = \\ = -[\Phi'(\mathbf{m}_1(t)) - \Phi'(\mathbf{m}_2(t))]w_2 - [\mathbb{P}_{\mathbf{m}_1(t)} - \mathbb{P}_{\mathbf{m}_2(t)}]w_2 := h_{\mathbf{m}_1,\mathbf{m}_2}.$$

Using now that $\Phi(u)$ is smooth and of order $2l-1$ together with estimates (4.26) and the θ-weighted analog of (5.15) (see Remark 5.9), we infer

$$(7.51) \qquad \|h_{\mathbf{m}_1,\mathbf{m}_2}\|_{L^p_{\mathrm{b},\theta}(\mathbb{R}^{n+1})} \le C\|w_2\|_{W^{(1,2l),p}_{\mathrm{b}}(\mathbb{R}^n)} \|\mathbf{m}_1 - \mathbf{m}_2\|_{L^\infty_\theta(\mathbb{R}^{n+1})}.$$

It remains to note that, due to Theorem 7.5, $\|w_2\|_{W^{(1,2l),p}_{\mathrm{b}}(\mathbb{R}^{n+1})} \le C\|h\|_{L^p_{\mathrm{b}}(\mathbb{R}^{n+1})}$ and, consequently, applying again Theorem 7.5 to equation (7.51), we obtain estimate (7.50) and finish the proof of Lemma 7.6. □

We now recall that the term $\mathbb{P}_{\mathbf{m}(t)} v$ in equation (7.7) is artificial and, therefore, we want to eliminate it. In order to do so, we impose the additional restriction $\mathbb{P}_{\mathbf{m}(t)} v(t) \equiv 0$, or, which is the same

$$(7.52) \qquad Z^i_r(t) := (v(t), \bar{\psi}^i_r(\mathbf{m}(t)) \equiv 0, \quad i = 1, \cdots, k, \quad r \in \mathbb{N}.$$

If $\mathbb{P}_{\mathbf{m}(t)} v(t) \equiv 0$ is satisfied, then,

$$\mathbb{P}_{\mathbf{m}(t)} \partial_t v(t) + \mathbb{D}(\mathbf{m}(t))[\mathbf{m}'(t)]v(t) \equiv 0$$

where the operator $\mathbb{D}(\mathbf{m}) = \mathbb{P}_{\mathbf{m}}\mathbb{P}'_{\mathbf{m}}$ is studied in Theorem 5.6. Applying now the operator $\mathbb{P}_{\mathbf{m}}$ to both sides of equation (7.7), and using the last relation, we infer

$$-\mathbb{D}(\mathbf{m}(t))[\mathbf{m}'(t)]v(t) + \mathbb{S}(\mathbf{m}(t))v(t) \equiv \mathbb{P}_{\mathbf{m}(t)}h(t) \tag{7.53}$$

with $\mathbb{S}(\mathbf{m})$ defined by 5.32. Thus, restriction $\mathbb{P}_{\mathbf{m}(t)}v(t) \equiv 0$ implies relation (7.53). The next proposition shows that this relation is also *sufficient* for $\mathbb{P}_{\mathbf{m}(t)}v(t) \equiv 0$ to be satisfied.

PROPOSITION 7.7. *Let $h \in L^p_{\mathrm{b}}(\mathbb{R}^{n+1})$ and $v \in W^{(1,2l),p}_{\mathrm{b}}(\mathbb{R}^{n+1})$ be the associated unique solution of equation (7.7). Assume also that (7.53) is satisfied for every $t \in \mathbb{R}$. Then, (7.52) is also satisfied for all $t \in \mathbb{R}$ and, consequently, v solves the following reduced form of equation (7.7):*

$$\partial_t v + A_0 v + \Phi'(\mathbf{m}(t))v = h. \tag{7.54}$$

PROOF. Indeed, applying operator $\mathbb{P}_{\mathbf{m}(t)}$ to both sides of equation (7.7) and using (7.53), we infer

$$\mathbb{P}_{\mathbf{m}(t)}\partial_t(\mathbb{P}_{\mathbf{m}(t)}v(t)) + \mathbb{P}_{\mathbf{m}(t)}v(t) = 0. \tag{7.55}$$

Using now that $\mathbb{P}_{\mathbf{m}(t)}v(t) = \sum_{j=1}^{\infty}\sum_{i=1}^{k} Z^i_j(t)\phi^i_{\Gamma_j(t)}$ and multiplying scalarly equation (7.55) by $\bar{\psi}^i_j(\mathbf{m})$, we will have

$$\frac{d}{dt}Z^i_j(t) + Z^i_j(t) + \mathbb{Q}^i_j(t)\mathbb{Z}(t) = 0 \tag{7.56}$$

where

$$\mathbb{Q}^i_j(t)\mathbb{Z}(t) := \sum_{l=1}^{\infty}\sum_{r=1}^{k} Z^r_l(t)(D_\Gamma\phi^i_{\Gamma_l}\Gamma'_l(t), \bar{\psi}^r_j(\mathbf{m}(t)))$$

and $\mathbb{Z}(t) = \{Z^r_j(t)\}_{j\in\mathbb{N}, i=1,\cdots,k} \in l^\infty$. Moreover, using estimates (2.32) and (5.1) and the fact that $\mathbf{m}'(t)$ is uniformly small, see (7.31) and arguing in a standard way, we have

$$\|\mathbb{Q}(t)\|_{\mathcal{L}(l^\infty, l^\infty)} \le C\nu \tag{7.57}$$

where the constant C is independent of L and of the concrete choice of the trajectory $\mathbf{m}(t) \in \mathbb{P}(L)$. This estimate shows that equation (7.56) has the form

$$\frac{d}{dt}\mathbb{Z}(t) + \mathbb{Z}(t) = \mathbb{Q}(t)\mathbb{Z}(t) \tag{7.58}$$

which is a small linear perturbation of the simplest one: $\frac{d}{dt}\mathbb{Z}(t) + \mathbb{Z}(t) = 0$. Since this equation, obviously, possesses an exponential dichotomy, then the perturbed equation possesses the same property if ν is small enough. It remains to note that, by the assumptions of the proposition, $v \in W^{(1,2l),p}_{\mathrm{b}}(\mathbb{R}^{n+1})$ and, consequently, $\mathbb{Z}(t)$ is, a priori, uniformly bounded as $t \to \infty$. Thus. $\mathbb{Z}(t) \equiv 0$ and Proposition 7.9 is proven. □

REMARK 7.8. Proposition 7.7 remains true also for *local* solutions v of (7.7) defined for $t \in [\tau, T]$ if it is known, in addition, that $\mathbb{P}_{\mathbf{m}(t_0)}v(t_0) = 0$ for some $t_0 \in [\tau, T]$. Indeed, the last assertion implies that $\mathbb{Z}(t_0) = 0$ and, due to the uniqueness theorem for equation (7.56), we have $\mathbb{Z}(t) \equiv 0$, $t \in [\tau, T]$.

We conclude this section by formulating the analogous results for the initial value problem

(7.59) $$\partial_t v + A_0 v + \Phi'(\mathbf{m}(t))v + \mathbb{P}_{\mathbf{m}(t)}v = 0, \quad t \geq \tau, \quad v\big|_{t=\tau} = v_\tau$$

under the additional assumption that the single pulse $V(x)$ is spectrally stable.

PROPOSITION 7.9. *Let the assumptions of Theorem 7.5 hold and, in addition, assume that the pulse V is spectrally stable. Then, there exists a positive constant β such that, for every weight function $\theta \in C(\mathbb{R}^n)$, every $\tau \in \mathbb{R}$ and every $v_\tau \in W_\theta^{(1,2l),p}(\mathbb{R}^n)$, problem (7.59) possesses a unique solution v and the following estimate holds:*

(7.60) $$\|v\|_{W_\theta^{(1,2l),p}([T,T+1]\times\mathbb{R}^n)} \leq C\,\mathrm{e}^{-\beta(T-\tau)}\,\|v_\tau\|_{W_\theta^{2l(1-1/p),p}(\mathbb{R}^n)}, \quad T \geq \tau$$

where the constant C depends on C_θ, but is independent of the concrete choice of θ, $\mathbf{m}(t)$ and v_τ. Moreover, the analogous result holds also for the spaces $W_{b,\theta}^{l,p}$. Furthermore, if equality (7.53) holds for every $t \geq \tau$ and $\mathbb{P}_{\mathbf{m}(\tau)}v_\tau = 0$, then

(7.61) $$\mathbb{P}_{\mathbf{m}(t)}v(t) \equiv 0, \quad \forall t \geq \tau.$$

Indeed, estimate (7.60) can be verified analogously to the proof of Theorems 7.5 (we only need to use, in addition, Corollary 3.11 in order to estimate the functions v_j from the proof of Theorem 7.5 on a halfline $t \in [\tau, \infty)$). In order to verify (7.61), it is sufficient to note that, under the assumptions of the proposition, the function $\mathbb{Z}(t)$ defined by (7.52) satisfies equation (7.58) for $t \geq \tau$ with the initial data $\mathbb{Z}(\tau) = 0$. Since the solution of (7.58) is unique, then necessarily $\mathbb{Z}(t) \equiv 0$ which implies (7.61) and finishes the proof of Proposition 7.9.

8. Slow evolution of multi-pulse structures: center manifold reduction

In this section, we construct a center manifold reduction for the following perturbed version of equation (2.1):

(8.1) $$\partial_t u + A_0 u + \Phi(u) = \mu R(t, u)$$

in the neighborhood of the multi-pulse manifold $\mathbb{P}(L)$ with sufficiently large L. Here $\mu \geq 0$ is a small parameter and $R(t,u) := R(t, x, u, D_x u, \cdots, D_x^{2l-1}u)$ is a smooth function with respect to u, $D_x u, \cdots, D_x^{2l-1}u$ and is uniformly bounded with respect to $(t, x) \in \mathbb{R}^{n+1}$.

To this end, we first fix the exponent $p = p(n)$ in such way that

(8.2) $$W_{\mathrm{b}}^{2l(1-1/p),p}(\mathbb{R}^n) \subset C_{\mathrm{b}}^{2l-1}(\mathbb{R}^n).$$

Then, as known (see e.g. [**Ama95**]), the Cauchy problem

(8.3) $$\partial_t u + A_0 u + \Phi(u) = \mu R(t, u), \quad u\big|_{t=\tau} = u_\tau$$

is locally uniquely solvable for every $u_\tau \in W_{\mathrm{b}}^{2l(1-1/p),p}(\mathbb{R}^n)$ and every $\tau \in \mathbb{R}$ and, thus, defines a local dynamical process $U(t,\tau)$ in the phase space $\mathbb{X}_{\mathrm{b}} := W_{\mathrm{b}}^{2l(1-1/p),p}(\mathbb{R}^n)$

(8.4) $$U(t,\tau)u_\tau := u(t), \quad u \text{ solves (8.3)}, \ \tau \leq t \leq t(u_\tau) > \tau.$$

We restrict ourselves to consider equation (8.1) in a sufficiently small neighborhood $\mathcal{O}_\kappa(\mathbb{P}(L))$ of the multi-pulse manifold $\mathbb{P}(L)$. According to Theorem 7.1, every such solution $u(t)$ can be uniquely decomposed as follows:

$$u(t) = \mathbf{m}(t) + v(t), \quad \mathbf{m}(t) \in \mathbb{P}(L), \quad \mathbb{P}_{\mathbf{m}(t)}v(t) \equiv 0. \tag{8.5}$$

Inserting now expression (8.5) into equation (8.1) and using that $A_0\mathbf{m} = \mathbb{F}(\mathbf{m}) - \Phi(\mathbf{m})$, see (4.29), we obtain the following equivalent form of equation (8.1):

$$\partial_t v + A_0 v + \Phi'(\mathbf{m}(t))v = -\mathbb{F}(\mathbf{m}(t)) - \Phi(v(t), \mathbf{m}(t)) + \mu R(t, v(t) + \mathbf{m}(t)) - \mathbf{m}'(t) \tag{8.6}$$

where $\Phi(v, \mathbf{m}) := \Phi(v + \mathbf{m}) - \Phi(\mathbf{m}) - \Phi'(\mathbf{m})v$. Furthermore, since we are interested in the small neighborhood of $\mathbb{P}(L)$ and (consequently) the rate $\mathbf{m}'(t)$ of pulse evolution is also expected to be small, we assume from now on that

$$\|v(t)\|_{W_b^{2l(1-1/p),p}(\mathbb{R}^n)} + \|\mathbf{m}'(t)\|_{L^\infty(\mathbb{R}^n)} \le \kappa \tag{8.7}$$

where κ is small enough that all of the assumptions of previous sections are satisfied (this inequality will be justified below).

We now recall that $\mathbb{P}_{\mathbf{m}(t)}v(t) \equiv 0$ and, consequently, due to (7.53),

$$\begin{aligned}(\mathrm{Id} - \mathbb{D}(\mathbf{m}(t))[\cdot]v(t))\mathbf{m}'(t) &= \\ = \mathbb{P}_{\mathbf{m}(t)}(-\mathbb{F}(\mathbf{m}(t)) &- \Phi(v(t), \mathbf{m}(t)) + \mu R(t, v(t) + \mathbf{m}(t))) - \mathbb{S}(\mathbf{m}(t))v(t)\end{aligned} \tag{8.8}$$

which will be interpreted as a differential equation for $\mathbf{m}$ on the manifold $\mathbb{P}(L)$. To this end, we need the following lemma, which allows us to solve it with respect to $\mathbf{m}'(t)$.

LEMMA 8.1. *Let the above assumptions hold and let $v \in L^\infty(\mathbb{R}^n)$ be such that $\|v\|_{L^\infty(\mathbb{R}^n)} \le \kappa$ with κ being small enough. Then, for every $\mathbf{m} \in \mathbb{P}(L)$, the operator*

$$\mathbb{M}(\mathbf{m}, v)\,\mathrm{w} := (\mathrm{Id} - \mathbb{D}(\mathbf{m})[\cdot]v)^{-1}\,\mathrm{w}$$

is well defined as an operator from $L^\infty(\mathbb{R}^n)$ to $\mathbb{P}(L)$, depends smoothly on $\mathbf{m}$ and v and its norms is uniformly bounded:

$$\|\mathbb{M}(\cdot, \cdot)\|_{C^k(\mathbb{P}(L)\times\{\|v\|_{L^\infty(\mathbb{R}^n)}<\kappa\})} \le C_k \tag{8.9}$$

for all $k \in \mathbb{N}$. Moreover, this operator is also Lipschitz continuous in weighted norms, i.e. for every $\mathbf{m}_i \in \mathbb{P}(L)$ $v_i \in L^\infty(\mathbb{R}^n)$ such that $\|v_i\|_{L^\infty(\mathbb{R}^n)} < \kappa$ and $\mathrm{w}_i \in T_{\mathbf{m}}\mathbb{P}(L)$, we have

$$\begin{aligned}\|\mathbb{M}(\mathbf{m}_1, v_1)\,\mathrm{w}_1 - \mathbb{M}(\mathbf{m}_2, v_2)\,\mathrm{w}_2\|_{L^\infty_{e^{-\gamma|x|}}(\mathbb{R}^n)} &\le C(\|\mathrm{w}_1 - \mathrm{w}_2\|_{L^\infty_{e^{-\gamma|x|}}(\mathbb{R}^n)} + \\ + (\|\mathrm{w}_1\|_{L^\infty(\mathbb{R}^n)} + \|\mathrm{w}_2\|_{L^\infty(\mathbb{R}^n)})(\|\mathbf{m}_1 - \mathbf{m}_2\|_{L^\infty_{e^{-\gamma|x|}}(\mathbb{R}^n)} &+ \|v_1 - v_2\|_{L^\infty_{e^{-\gamma|x|}}(\mathbb{R}^n)}))\end{aligned} \tag{8.10}$$

where the constant C is independent of L, $\mathbf{m}_i$, v_i and w_i and $\gamma > 0$ is small enough.

Indeed, all of the assertions of the lemma follow immediately from Theorem 5.6 and the standard presentation of the operator $\mathbb{M}$ as the Neumann series

$$\mathbb{M}(\mathbf{m}, v) = \sum_{i=1}^{\infty} (\mathbb{D}(\mathbf{m})[\cdot]v)^i.$$

Thus, equation (8.8) can be rewritten in the following more convenient form:

$$\mathbf{m}'(t) = \mathrm{f}(t, \mathbf{m}(t), v(t)) \tag{8.11}$$

where

$$\mathrm{f}(t, \mathbf{m}, v) := \mathbb{M}(\mathbf{m}, v)\big(-\mathbb{P}_{\mathbf{m}}(\mathbb{F}(\mathbf{m}) - \Phi(v, \mathbf{m}) + \mu R(t, v + \mathbf{m})) - \mathbb{S}(\mathbf{m})v\big). \tag{8.12}$$

The next lemma collects the main properties of the function f which are factually already proven above.

LEMMA 8.2. *Let the above assumptions hold. Then the function* f *is uniformly smooth on* $\mathbb{P}(L) \times \{\|v\|_{W_b^{(2l(1-1/p),p}(\mathbb{R}^n)} < \kappa\}$ *for every fixed* t *and, in particular, the following estimates hold:*
(8.13)

$$\begin{cases} \| \mathrm{f}(t,\mathbf{m},v)\|_{L^\infty(\mathbb{R}^n)} + \| \mathrm{f}'_{\mathbf{m}}(t,\mathbf{m},v)\|_{\mathcal{L}(T_{\mathbf{m}}\mathbb{P}(L),L^\infty(\mathbb{R}^n))} \le C_\varepsilon(\mathrm{e}^{-2(\alpha-\varepsilon)L} + \mu + \kappa^2) \\ \| \mathrm{f}'_v(t,\mathbf{m},v)\|_{\mathcal{L}(\mathbb{X}_b(\mathbb{R}^n),L^\infty(\mathbb{R}^n))} \le C_\varepsilon(\mathrm{e}^{-2(\alpha-\varepsilon)L} + \mu + \kappa) \end{cases}$$

where $\varepsilon > 0$ *is arbitrary, and the constant* C_ε *is independent of* t, $\mathbf{m}$, v *and* $L \ge L_0(\varepsilon)$. *Moreover,* f *is also Lipschitz continuous in the local topology as well, namely*

$$(8.14)\quad \|\mathrm{f}(t,\mathbf{m}_1,v_1) - \mathrm{f}(t,\mathbf{m}_2,v_2)\|_{L^\infty_{\mathrm{e}^{-\gamma|x|}}(\mathbb{R}^n)} \le \\ \le C(\mathrm{e}^{-\alpha L} + \mu + \kappa)(\|\mathbf{m}_1 - \mathbf{m}_2\|_{L^\infty_{\mathrm{e}^{-\gamma|x|}}(\mathbb{R}^n)} + \|v_1 - v_2\|_{W^{2l(1-1/p),p}_{b,\mathrm{e}^{-\gamma|x|}}(\mathbb{R}^n)})$$

where $\gamma > 0$ *s small enough and* C *is independent of* t, $\mathbf{m}_i$, v_i *and* L.

Indeed, the required estimates for operators $\mathbb{F}(\mathbf{m})$, $\mathbb{S}(\mathbf{m})$, $\mathbb{P}_{\mathbf{m}}$ and $\mathbb{M}(\mathbf{m},v)$ are obtained in Corollary 4.10, Theorem 5.8, Theorem 5.2 and Lemma 8.1 respectively. Combining these estimates and taking into the account that $\Phi(\mathbf{m},0) = \Phi'_v(\mathbf{m},0) = 0$, we receive all of the estimates stated in the lemma.

Furthermore, in order to solve equation (8.6), we transform it to the form of (7.7) (taking into the account that $\mathbb{P}_{\mathbf{m}(t)}v(t) \equiv 0$)

$$(8.15)\qquad \partial_t v + A_0 v + \Phi(\mathbf{m}(t))v + \mathbb{P}_{\mathbf{m}(t)}v = h_\mu(t,\mathbf{m}(t),v(t),\mathbf{m}'(t))$$

with

$$(8.16)\qquad h_\mu(t,\mathbf{m},v,\mathrm{w}) := -\mathbb{F}(\mathbf{m}) - \Phi(v,\mathbf{m}) + \mu R(t,v+\mathbf{m}) - \mathrm{w}\,.$$

We recall that, by the construction, equation (8.11) is exactly condition (7.53) for equation (8.15) which, due to Proposition 7.7 and Remark 7.8, guarantees that the additional artificial term $\mathbb{P}_{\mathbf{m}(t)}v(t)$ vanishes identically. Thus, if $u(t)$, $t \in [\tau, T]$ solves (8.1) and $u(t) \in \mathcal{O}_\kappa(\mathbb{P}(L))$ on $[\tau, T]$ then the associated functions $\mathbf{m}(t)$ and $v(t)$ satisfies on $[\tau, T]$ the following system:

$$(8.17)\qquad \begin{cases} \partial_t v + A_0 v + \Phi(\mathbf{m}(t))v + \mathbb{P}_{\mathbf{m}(t)}v = h_\mu(t,\mathbf{m}(t),v(t),\mathbf{m}'(t)), \\ \mathbf{m}' = \mathrm{f}(t,\mathbf{m},v) \end{cases}$$

and, vise versa, any solution $(\mathbf{m}(t), v(t))$, $t \in [\tau, T]$ with sufficiently small v and satisfying $\mathbb{P}_{\mathbf{m}(\tau)}v(\tau) = 0$ generates a unique solution $u(t) := \mathbf{m}(t) + v(t)$ of equation (8.1). Therefore, instead of the initial equation (8.1), it is sufficient to solve the associated system (8.17).

Nevertheless system (8.17) is still inconvenient for the center manifold reduction. Indeed, we are going to use the cut-off operator $\mathrm{Cut}(\mathbf{m})$ in order to eliminate the influence of the boundary $\partial\mathbb{P}(L)$. However, the equation *does not coincide* with condition (7.53) for the first equation and, consequently, we cannot eliminate the artificial term $\mathbb{P}_{\mathbf{m}}v$.

In order to overcome this difficulty, we replace the function h in the first equation of (8.17) by the following one:

$$H(t,\mathbf{m},v,\mathbf{m}') = h_\mu - \mathbb{P}_{\mathbf{m}} h_\mu - \mathbb{D}(\mathbf{m})[\mathbf{m}']v + \mathbb{S}(\mathbf{m})v = \\ = (\mathrm{Id} - \mathbb{P}_{\mathbf{m}})(-\mathbb{F}(\mathbf{m}) - \Phi(v,\mathbf{m}) + \mu R(t, v+\mathbf{m})) - \mathbb{D}(\mathbf{m})[\mathbf{m}']v + \mathbb{S}(\mathbf{m})v. \tag{8.18}$$

Indeed, on the one hand, if the second equation of (8.17) is satisfied we have $h_\mu = H$, see (8.8). Consequently, (8.17) is equivalent to the following one:

$$\begin{cases} \partial_t v + A_0 v + \Phi(\mathbf{m}(t))v + \mathbb{P}_{\mathbf{m}(t)} v = H(t, \mathbf{m}(t), v(t), \mathbf{m}'(t)), \\ \mathbf{m}' = \mathrm{f}(t, \mathbf{m}, v). \end{cases} \tag{8.19}$$

On the other hand, as is not difficult to verify, condition (7.53) is *automatically* satisfied if the first equation of (8.19) holds (independently of the validity of the second equation). Thus, the term $\mathbb{P}_{\mathbf{m}(t)} v(t)$ is now controllable and we can, indeed, cut-off the second equation. To this end, using the cut-off operator introduced in Theorem 6.6, we finally transform system (8.19) as follows

$$\begin{cases} \partial_t v + A_0 v + \Phi(\mathbf{m}(t))v + \mathbb{P}_{\mathbf{m}(t)} v = \mathbb{H}(t, \mathbf{m}(t), v(t)), \\ \mathbf{m}' = \widetilde{\mathrm{f}}(t, \mathbf{m}(t), v(t)) := \mathrm{Cut}(\mathbf{m}(t))\, \mathrm{f}(t, \mathbf{m}(t), v(t)) \end{cases} \tag{8.20}$$

with

$$\mathbb{H}(t,\mathbf{m},v) := H(t,\mathbf{m},v,\widetilde{\mathrm{f}}(t,\mathbf{m},v)). \tag{8.21}$$

Indeed, equations (8.19) and (8.20) are no more equivalent, but, due to the construction of the cut-off operator, they coincide for all $\mathbf{m} \in \mathbb{P}(L')$ where

$$L' > (1+\varepsilon)L$$

(and $\varepsilon > 0$ tends to zero as $L \to \infty$). Consequently, these equations are still equivalent as long as the trajectory $\mathbf{m}(t)$ remains inside of $\mathbb{P}(L')$ (and $v(t)$ remains inside of $\mathcal{O}_\delta(\mathbb{P}(L'))$). On the other hand, due to Theorem 6.6, the trajectory $\mathbf{m}(t)$ cannot now reach the boundary $\partial\mathbb{P}(L)$ in finite time and, therefore, we need not to take care on the "boundary effects".

Thus, instead of studying the initial equation (8.1), it is sufficient to investigate the more convenient system (8.20).

The next lemma collects, analogously to Lemma 8.2, the main properties of the function H which are factually verified in the previous sections.

LEMMA 8.3. *Let the above assumptions hold and let*

$$\|v\|_{W_b^{2l(1-1/p),p}(\mathbb{R}^n)} < \kappa, \tag{8.22}$$

for a sufficiently small κ. Then, the function $\mathbb{H}$ defined by (8.21) is uniformly smooth with respect to $\mathbf{m} \in \mathbb{P}(L)$ and v satisfying (8.22) and the following estimates hold:

$$\begin{cases} \|\mathbb{H}(t,\mathbf{m},v)\|_{L_b^p(\mathbb{R}^n)} + \\ \qquad \|\mathbb{H}'_{\mathbf{m}}(t,\mathbf{m},v)\|_{\mathcal{L}(T_{\mathbf{m}}\mathbb{P}(L), L_b^p(\mathbb{R}^n))} \le C_\varepsilon(\mathrm{e}^{-2(\alpha-\varepsilon)L} + \kappa^2 + \mu), \\ \|\mathbb{H}'_v(t,\mathbf{m},v)\|_{\mathcal{L}(W_b^{2l(1-1/p),p}(\mathbb{R}^n), L_b^p(\mathbb{R}^n))} \le C_\varepsilon(\mathrm{e}^{-2(\alpha-\varepsilon)L} + \mu + \kappa) \end{cases} \tag{8.23}$$

where $\varepsilon > 0$ is arbitrary and the constants C and C_ε are independent of L, t, $\mathbf{m}$ and v. Moreover, the weighted Lipschitz continuity also holds:

$$(8.24)\quad \|\mathbb{H}(t,\mathbf{m}_1,v_1)-\mathbb{H}(t,\mathbf{m}_2,v_2)\|_{L^p_{b,e^{-\gamma|x|}}(\mathbb{R}^n)} \le C\,e^{+\delta L}(e^{-\alpha L}+\mu+\kappa)\times \\ \times\left(\|\mathbf{m}_1-\mathbf{m}_2\|_{L^\infty_{e^{-\gamma|x|}}(\mathbb{R}^n)}+\|v_1-v_2\|_{W^{2l(1-1/p),p}_{b,e^{-\gamma|x|}}(\mathbb{R}^n)}\right)$$

for $\gamma > 0$ and $\delta > 0$ sufficiently small.

Indeed, as in the previous lemma, the required estimates for operators $\mathbb{F}(\mathbf{m})$, $\mathbb{S}(\mathbf{m})$, $\mathbb{P}_{\mathbf{m}}$, f and Cut are obtained in Corollary 4.10, Theorem 5.8, Theorem 5.2, Lemma 8.2 and Theorem 6.6 respectively. Combining these estimates and taking into account that $\Phi(\mathbf{m},0)=\Phi'_v(\mathbf{m},0)=0$, we receive all of the estimates stated in the lemma (we emphasize that the growing multiplier $e^{\delta L}$ comes from the analogous growing estimate for the operator Cut, see Theorem 6.6).

REMARK 8.4. As before, we have formulated in Lemma 8.2 and 8.3 the weighted estimates for the case of weights $e^{-\gamma|x|}$ only. In a fact, they holds for all weights $e^{-\gamma|x-x_0|}$ uniformly with respect to $x_0\in\mathbb{R}^n$.

Thus, Lemmas 8.2 and 8.3 together with Theorem 7.5 show that equations (8.20) give indeed the splitting of the initial problem (8.1) into slow ($\mathbf{m}$) and fast (v) variables. The next theorem, which establish the existence of a center manifold reduction for that system is the main result of the section.

THEOREM 8.5. *Let the assumptions of Section 2 hold. Then, for every finite $k\in\mathbb{N}$ and every $\varepsilon>0$, there exist $L_0=L_0(k,\varepsilon)$ and $\mu_0=\mu_0(k)$ such that, for every $L>L_0$ and $|\mu|\le\mu_0$, there is a unique map*

$$\mathbb{W}:\mathbb{R}\times\mathbb{P}(L)\to W_b^{2l(1-1/p),p}(\mathbb{R}^n)$$

which is C^k-smooth with respect to $\mathbf{m}\in\mathbb{P}(L)$ and satisfies

$$(8.25)\qquad \|\mathbb{W}(t,\cdot)\|_{C^k(\mathbb{P}(L),W_b^{2l(1-1/p),p}(\mathbb{R}^n))}\le C_\varepsilon(e^{-2(\alpha-\varepsilon)L}+\mu)$$

where C_ε is independent of t, L and μ. This map possesses the following properties:

1) For every $\tau\in\mathbb{R}$ and every $\mathbf{m}_\tau\in\mathbb{P}(L)$ the solution $\mathbf{m}(t)$ of the equation

$$(8.26)\qquad \frac{d}{dt}\mathbf{m}(t)=\tilde{\mathrm{f}}(t,\mathbf{m}(t),\mathbb{W}(t,\mathbf{m}(t))),\quad \mathbf{m}(\tau)=\mathbf{m}_\tau$$

is globally defined for all $t\in\mathbb{R}$ and generates an associated solution of problem (8.20) via

$$v(t):=\mathbb{W}(t,\mathbf{m}(t)).$$

2) Vise versa, every solution $(\mathbf{m}(t),v(t))$ of problem (8.20) which is defined for all $t\in\mathbb{R}$ and whose v component belongs to a sufficiently small neighborhood of $\mathbb{P}(L)$ for all t can be represented in the form $(\mathbf{m}(t),\mathbb{W}(t,\mathbf{m}(t)))$ for some solution $\mathbf{m}(t)$ of (8.26).

3) For all $t\in\mathbb{R}$ and all $\mathbf{m}\in\mathbb{P}(L)$, we have

$$\mathbb{P}_{\mathbf{m}}\mathbb{W}(t,\mathbf{m})\equiv 0.$$

Moreover, if, in addition, the quantity $e^{2\gamma L}(e^{-\alpha L}+\mu)$ is small enough, the map $\mathbb{W}(t,\cdot)$ is also uniformly Lipschitz continuous in the local topology, i.e.

$$(8.27)\quad \|\mathbb{W}(t,\mathbf{m}_1)-\mathbb{W}(t,\mathbf{m}_2)\|_{W^{2l(1-1/p),p}_{b,e^{-\gamma|x|}}(\mathbb{R}^n)}\le C(\mu+e^{-\alpha L})\|\mathbf{m}_1-\mathbf{m}_2\|_{L^\infty_{e^{-\gamma|x|}}(\mathbb{R}^n)}$$

where $\gamma > 0$ is small enough.

REMARK 8.6. Theorem 8.5 claims that the set

$$\widetilde{\mathbb{Q}} := \{(t, \mathbf{m}, \mathbb{W}(t, \mathbf{m})),\ t \in \mathbb{R},\ \mathbf{m} \in \mathbb{P}(L)\}$$

is a globally invariant (center) manifold for (8.20) in the extended phase space $\mathbb{R} \times \mathbb{P}(L) \times \mathbb{X}_b$. This manifold generates a family of manifolds in the phase space $\mathbb{X}_{\mathrm{b}}$ of the initial problem (8.1) via

$$\mathbb{U}_L(t) := \{u = \mathbf{m} + \mathbb{W}(t, \mathbf{m}),\quad \mathbf{m} \in \mathbb{P}(L)\}.$$

Indeed, estimate (8.25) (together with the orthogonality of $\mathbb{W}$) guarantees that they are C^k-submanifolds of $\mathbb{X}_{\mathrm{b}}$ globally diffeomorphic to $\mathbb{P}(L)$. However, these manifolds are not invariant with respect to the flow generated by equation (8.1), since the construction of equations (8.20) involves the cut-off procedure. Nevertheless, if we restrict ourselves to consider slightly smaller manifolds

$$\mathbb{U}_{L'}(t) := \{u = \mathbf{m} + \mathbb{W}(t, \mathbf{m}),\quad \mathbf{m} \in \mathbb{P}(L')\} \subset \mathbb{U}_L(t)$$

with $L' = (1 + \varepsilon)L$, the time dependent family $\mathbb{U}' = \mathbb{U}'(t)$ of manifolds will be locally invariant with respect to the evolution governed by (8.1), i.e., the trajectory $u(t)$ of equation (8.1) can enter or go out from $\mathbb{U}_{L'}$ only through the boundary $\partial\mathbb{U}_{L'}$ and the dynamics on the family $\mathbb{U}_{L'}$ is governed by the reduced equations (8.26). Thus, the time-dependent family $\mathbb{U}_{L'}(t)$ gives, indeed, the non-autonomous center manifold reduction for (8.1) near the multi-pulse manifold $\mathbb{P}(L')$.

Beginning of Proof of Theorem 8.5. As usual, we restrict ourselves by constructing only the *Lipschitz* continuous center manifold. The smoothness of function $\mathbb{W}$ can be proven in a standard way using e.g. the fiber contraction arguments, see the monograph [**SSTC01**] or [**Mie86, Sak90, Mie88, VaV87**].

According to the general scheme, and using the Banach contraction principle, we are going to check that, for any $\tau \in \mathbb{R}$ and $\mathbf{m}_\tau \in \mathbb{P}(L)$, system (8.20) possesses a unique solution $(\mathbf{m}(t), v(t))$, $t \in \mathbb{R}$, such that $\mathbf{m}(\tau) = \mathbf{m}_\tau$ and $v(t) \in \mathcal{O}(\mathbb{P}(L))$ for all $t \in \mathbb{R}$. Then, we set $\mathbb{W}(\tau, \mathbf{m}_\tau) := v(\tau)$ and obtain the required center manifold.

Let us consider now a ball

$$\mathcal{B}_\kappa := \{v \in W_{\mathrm{b}}^{(1,2l),p}(\mathbb{R}^{n+1}),\ \|v\|_{W_{\mathrm{b}}^{(1,2l),p}(\mathbb{R}^{n+1})} \le \kappa\}$$

where κ is a sufficiently small positive number, which will be fixed below. Then, due to Theorems 6.1 and 6.6 and Lemma 8.2, for every $\tau \in \mathbb{R}$, every $\mathbf{m}_\tau \in \mathbb{P}(L)$ and every $v \in \mathcal{B}_\kappa$, the second equation of (8.20):

$$\mathbf{m}'(t) = \widetilde{\mathrm{f}}(t, \mathbf{m}(t), v(t)),\quad \mathbf{m}(\tau) = \mathbf{m}_\tau$$

possesses a unique solution $\mathbf{m} = \mathcal{S}_\tau(\mathbf{m}_\tau, v)$. Moreover, due to Lemma 8.2 and Theorem 6.6, we have

$$\|\mathbf{m}'(t)\|_{L^\infty(\mathbb{R}^n)} \le C_\varepsilon(\mathrm{e}^{-2(\alpha-\varepsilon)L} + \kappa^2 + \mu) \tag{8.28}$$

where C_ε depends on $\varepsilon > 0$, but is independent of $L > L_0(\varepsilon)$, μ, τ and $\mathbf{m}_\tau$. Thus, the rate of pulse evolution is indeed slow (e.g. assumption (7.31) is satisfied) and we can apply Theorem 7.5 in order to invert the linear part of the first equation (8.20). To be more precise, we define the following operator:

$$\mathbb{L}(\tau, \mathbf{m}_\tau, v) := \mathbb{T}_{\mathcal{S}_\tau(\mathbf{m}_\tau, v)}\mathbb{H}(\cdot, \mathcal{S}_\tau(\mathbf{m}_\tau, v), v(\cdot)) \tag{8.29}$$

where $\mathbb{T}_{\mathbf{m}}$ is the solution operator defined by (7.48). Then, finding bounded solutions of (8.20) is equivalent to finding fixed points of the operator (8.29). The next Lemma shows that this map is really well defined on $\mathcal{B}_\kappa$.

LEMMA 8.7. *Let the above assumptions hold and let*

$$\kappa_0 := C_\varepsilon(\mathrm{e}^{-2(\alpha-\varepsilon)L} + \mu) \tag{8.30}$$

for the appropriate constant C_ε depending only on ε. Then, the map $\mathbb{L}$ defined above maps $\mathcal{B}_\kappa$ into itself

$$\mathbb{L}(\tau, \mathbf{m}_\tau, \cdot) : \mathcal{B}_{\kappa_0} \to \mathcal{B}_{\kappa_0}, \quad \forall \tau \in \mathbb{R}, \quad \mathbf{m}_\tau \in \mathbb{P}(L)$$

if $L > L_0(\varepsilon)$, $|\mu| < \mu_0(\varepsilon)$.

PROOF. Indeed, according to Lemma 8.3 and Theorem 7.5, we have

$$\|\mathbb{L}(\tau, \mathbf{m}_\tau, v)\|_{W_{\mathrm{b}}^{(1,2l),p}(\mathbb{R}^{n+1})} \le C'_\varepsilon(\mathrm{e}^{-2(\alpha-\varepsilon)L} + \kappa^2 + \mu)$$

if $v \in \mathcal{B}_\kappa$. So, we only need to find κ satisfying

$$\kappa \le C'_\varepsilon(\mathrm{e}^{-2(\alpha-\varepsilon)L} + \kappa^2 + \mu).$$

It is not difficult to see that, for sufficiently large L and small μ, this inequality has a solution of the form (8.30) and Lemma 8.7 is proven. $\square$

Our next aim is to verify that $\mathbb{L}$ is a contraction in the appropriate metric weighted in time.

LEMMA 8.8. *Let the above assumptions hold. Then, for sufficiently large L and small μ, there exists a positive constant γ (independent of L and μ) such that*

$$\begin{aligned} &\|\mathbb{L}(\tau, \mathbf{m}_\tau^1, v_1) - \mathbb{L}(\tau, \mathbf{m}_\tau^2, v_2)\|_{W^{(1,2),p}_{b,\mathrm{e}^{-\gamma|t-\tau|}}(\mathbb{R}^{n+1})} \le \\ &\le C_\varepsilon(\mathrm{e}^{-2(\alpha-\varepsilon)L} + \mu)(\|\mathbf{m}_\tau^1 - \mathbf{m}_\tau^2\|_{L^\infty(\mathbb{R}^{n+1})} + \|v_1 - v_2\|_{W^{(1,2),p}_{b,\mathrm{e}^{-\gamma|t-\tau|}}(\mathbb{R}^{n+1})}) \end{aligned} \tag{8.31}$$

uniformly with respect to $\tau \in \mathbb{R}$, $\mathbf{m}_\tau^i \in \mathbb{P}(L)$ and $v_i \in \mathcal{B}_{\kappa_0}$.

PROOF. Indeed, due to Lemma 8.2 and Theorem 6.6, we have

$$\begin{aligned} &\|\widetilde{\mathrm{f}}(t, \mathbf{m}_1, v_1) - \widetilde{\mathrm{f}}(t, \mathbf{m}_2, v_2)\|_{L^\infty(\mathbb{R}^n)} \le \\ &\qquad \le \kappa'_0(\|\mathbf{m}_1 - \mathbf{m}_2\|_{L^\infty(\mathbb{R}^n)} + \|v_1(t) - v_2(t)\|_{W_{\mathrm{b}}^{2l(1-1/p),p}(\mathbb{R}^n)}) \end{aligned}$$

with $\kappa'_0 := C\kappa_0$. Consequently, denoting $\mathbf{m}_i := \mathcal{S}(\tau, \mathbf{m}_\tau^i, v_i)$ and applying estimate (6.4), we get

$$\begin{aligned} &\mathrm{e}^{-\kappa'|t-\tau|}\|\mathbf{m}_1(t) - \mathbf{m}_2(t)\|_{L^\infty(\mathbb{R}^n)} \le \\ &\le \|\mathbf{m}_\tau^1 - \mathbf{m}_\tau^2\|_{L^\infty(\mathbb{R}^n)} + \kappa' \operatorname{sgn}(t-\tau) \int_\tau^t \mathrm{e}^{-\kappa'|s-\tau|}\|v_1(s) - v_2(s)\|_{W_{\mathrm{b}}^{2l(1-1/p),p}(\mathbb{R}^n)}\, ds. \end{aligned}$$

Multiplying this inequality by $e^{-(\gamma-\kappa')|t-\tau|}$ for some $\gamma > \kappa'$ and taking the supremum over $t \in \mathbb{R}$, after the standard estimates, we infer

$$(8.32)\quad \|\mathbf{m}_1 - \mathbf{m}_2\|_{L^\infty_{e^{-\gamma|t-\tau|}}(\mathbb{R}^{n+1})} \le \|\mathbf{m}^1_\tau - \mathbf{m}^2_\tau\|_{L^\infty(\mathbb{R}^n)} + \\ + \frac{\kappa'}{\gamma-\kappa'}\sup_{t\in\mathbb{R}}\{e^{-\gamma|s-\tau|}\,\|v_1(s) - v_2(s)\|_{W^{2l(1-1/p),p}_{\mathrm{b}}(\mathbb{R}^n)}\} \le \|\mathbf{m}^1_\tau - \mathbf{m}^2_\tau\|_{L^\infty(\mathbb{R}^n)} + \\ + C\frac{\kappa'}{\gamma-\kappa'}\|v_1 - v_2\|_{W^{(1,2l),p}_{b,e^{-\gamma|t-\tau|}}(\mathbb{R}^{n+1})}.$$

Using now this formula together with the fact that $v_i \in \mathcal{B}_{\kappa_0}$ and Lemma 8.3, we obtain the analogous estimate for the function $\mathbb{H}$:

$$\|\mathbb{H}(\cdot,\mathbf{m}_1,v_1) - \mathbb{H}(\cdot,\mathbf{m}_2,v_2)\|_{L^p_{b,e^{-\gamma|t-\tau|}}(\mathbb{R}^{n+1})} \le \\ \le C\kappa_0(\|\mathbf{m}^1_\tau - \mathbf{m}^2_\tau\|_{L^\infty(\mathbb{R}^n)} + \|v_1 - v_2\|_{W^{(1,2l),p}_{b,e^{-\gamma|t-\tau|}}(\mathbb{R}^{n+1})})$$

we have assumed, for simplicity, that $\gamma > 2\kappa_0$.

Finally, fixing $\gamma > 2\kappa_0$ small enough such that Theorem 7.5 and Lemma 7.6 hold for the weight $e^{-\gamma|t-\tau|}$, we deduce the required estimate (8.31) (we note that the validity of Theorem 7.5 and Lemma 4.4 requires $\gamma < \varepsilon_0$ *independent* of L and μ, so such γ exists for sufficiently small κ_0). Lemma 8.8 is proven. □

End of Proof of Theorem 8.5. It is now not difficult to finish the proof of the theorem. Indeed, according to Lemmas 8.7 and 8.11, for every fixed $\tau \in \mathbb{R}$ and $\mathbf{m}_\tau \in \mathbb{P}(L)$, the map $\mathbb{L}(\tau,\mathbf{m}_\tau,\cdot)$ is a contraction on $\mathcal{B}_\tau$ (endowed by the topology of $W^{(1,2l),p}_{b,e^{-\gamma|t-\tau|}}(\mathbb{R}^{n+1})$), if $\kappa_0(L,\mu)$ is small enough. Since, $\mathcal{B}_\tau$ endowed by this topology is, obviously, a complete metric space, then, due to the Banach contraction principle, this map has a unique fixed point $v = V(\tau,\mathbf{m}_\tau) \in \mathcal{B}_{\kappa_0}$. Moreover, since this map is Lipschitz continuous also in $\mathbf{m}_\tau$, we have

$$(8.33)\qquad \|V(\tau,\mathbf{m}^1_\tau) - V(\tau,\mathbf{m}^2_\tau)\|_{W^{(1,2l),p}_{b,e^{-\gamma|t-\tau|}}(\mathbb{R}^{n+1})} \le C\kappa_0\|\mathbf{m}^1_\tau - \mathbf{m}^2_\tau\|_{L^\infty(\mathbb{R}^n)}.$$

We now set

$$\mathbb{W}(\tau,\mathbf{m}_\tau) := V(\tau,\mathbf{m}_\tau)(\tau).$$

We claim that the map $\mathbb{W}$ thus defined satisfies all assertions of the theorem. Indeed, the Lipschitz analog of (8.25) is an immediate corollary of (8.33).

Let now $(\mathbf{m}(t), v(t))$ be a bounded solution of (8.20) such that $v \in \mathcal{B}_{\kappa_0}$. Then, due to the uniqueness of $\mathbb{W}$, we have

$$v(t) = \mathbb{W}(t,\mathbf{m}(t)),\quad t\in\mathbb{R}$$

and, consequently, $\mathbf{m}(t)$ solves the reduced problem (8.26). Vice versa, the fact that any solution $\mathbf{m}(t)$ of the reduced problem generates the associated solution of problem (8.20) is an immediate corollary of our construction of map $\mathbb{L}$ and the property $\mathbb{P}_\mathbf{m}\mathbb{W}(t,\mathbf{m}) \equiv 0$ follows from the fact, that, due to Proposition 7.7, any solution $(\mathbf{m}(t), v(t))$ of (8.20) which is defined for all t and bounded, necessarily satisfies $\mathbb{P}_{\mathbf{m}(t)}v(t) \equiv 0$.

So, it only remains to verify the weighted Lipschitz continuity (8.27). To this end, it is sufficient to verify that the operator $\mathbb{L}$ is a contraction in the space-time weighted metric of $W^{(1,2l),p}_{b,e^{-\gamma|t-\tau|-\gamma|x|}}(\mathbb{R}^{n+1})$ as well. Indeed, since all of the "space-uniform" estimates involved into the proof of Lemma 8.8 have their space weighted analogs, we also have the space-weighted analog of estimate (8.31), but with the Lipschitz constant $C\,\mathrm{e}^{\delta L}(\mathrm{e}^{-\alpha L}+\mu)$. If this quantity is small enough, we have the contraction in the space-time weighted norm which, in turns, gives (8.27). Theorem 8.5 is proven. □Let us now study the dependence of the function $\mathbb{W}$ on t. To this end, it is however more convenient to investigate the dependence of equation (8.1) on the perturbation R. To this end, we introduce a metric on the space of such perturbations via the following natural expression:

$$\|R_1(t,\cdot)-R_2(t,\cdot)\|_{C^m} := \|R_1(t,\cdot)-R_2(t,\cdot)\|_{C^m(O_\delta(\mathbb{P}(L)),L^\infty(\mathbb{R}^n))},$$

$m=0,1,\cdots$. Then, the following result holds.

COROLLARY 8.9. *Let the above assumptions hold and let $\mathbb{W}_{R_1}$ and $\mathbb{W}_{R_2}$ define the center manifolds for equation* (8.1) *with right-hand sides R_1 and R_2 respectively. Then, for any m, sufficiently large L and small μ, one has*

$$\text{(8.34)}\quad \|\mathbb{W}_{R_1}(t,\cdot)-\mathbb{W}_{R_2}(t,\cdot)\|_{C^m(\mathbb{P}(L),W_{\mathrm{b}}^{2l(1-1/p),p}(\mathbb{R}^n))}\le \\ \le C\mu\sup_{s\in\mathbb{R}}\{\mathrm{e}^{-\gamma|t-s|}\,\|R_1(s,\cdot)-R_2(s,\cdot)\|_{C^m}\}$$

where the positive constants C and γ depend on the C^m-norms of R_1 and R_2, but are independent of L, μ, t and s.

Indeed, including the dependence on R into the operator $\mathbb{L}$ and arguing exactly as in the proof of Lemma 8.8, we obtain the additional term of the form $C\mu\sup_{t\in\mathbb{R}}\{\mathrm{e}^{-\gamma|t-\tau|}\,\|R_1(t,\cdot)-R_2(t,\cdot)\|_{C^0}\}$ which,in turns gives (8.34) for $m=0$. The proof of estimate (8.34) for $m>0$ is analogous to the standard proof of the higher regularity of the center manifold and we leave it to the reader.

COROLLARY 8.10. *Let the above assumptions hold and let $\mathbb{W}(t,\cdot)$ define the center manifold for equation* (8.20) *with some fixed right-hand side R. Then, the following estimate holds:*

$$\text{(8.35)}\quad \|\mathbb{W}(t,\cdot)-\mathbb{W}(t+\tau,\cdot)\|_{C^m(\mathbb{P}(L),W_{\mathrm{b}}^{2l(1-1/p),p}(\mathbb{R}^n))}\le \\ \le C\mu\sup_{s\in\mathbb{R}}\{\mathrm{e}^{-\gamma|t-s|}\,\|R(s,\cdot)-R(s+\tau,\cdot)\|_{C^m}\}.$$

In particular, if $R(t,\cdot)$ is autonomous, time-periodic or almost-periodic in time, the same will be true for the manifold $\mathbb{W}$.

Indeed, in order to deduce (8.35), it is sufficient to apply estimate (8.34) with $R_1(t,\cdot)=R(t,\cdot)$ and $R_2(t,\cdot)=R(t+\tau,\cdot)$.

We now recall that the group of spatial symmetries G acts on the phase space of (8.1) by isometries $\mathcal{T}_\Gamma$, $\Gamma\in G$. Moreover, as it is not difficult to see, the pulse manifold $\mathbb{P}(L)$ and the projectors $\mathbb{P}_{\mathbf{m}}$ are invariant with respect to this action:

$$\mathcal{T}_\Gamma\mathbb{P}(L)=\mathbb{P}(L),\quad \mathbb{P}_{\mathcal{T}_\Gamma\mathbf{m}}=\mathcal{T}_\Gamma\circ\mathbb{P}_{\mathbf{m}}\circ\mathcal{T}_\Gamma^{-1}.$$

The next corollary shows that the map $\mathbb{W}(t,\cdot)$ is also invariant with respect to the part of group G which leaves invariant the perturbation R. This result will be used in our application to space-time chaos construction in the Swift-Hohenberg equations.

COROLLARY 8.11. *Let the above assumptions hold and let R be invariant with respect to some $\mathcal{T}_{\Gamma_0}$, $\Gamma_0 \in G$:*

$$\mathcal{T}_{\Gamma_0} R(t,u) \equiv R(t, \mathcal{T}_{\Gamma_0} u). \tag{8.36}$$

Then, the manifold $\mathbb{W}$ is also invariant with respect to $\mathcal{T}_{\Gamma_0}$:

$$\mathcal{T}_{\Gamma_0} \mathbb{W}(t,\mathbf{m}) = \mathbb{W}(t, \mathcal{T}_{\Gamma_0}\mathbf{m}). \tag{8.37}$$

Indeed, under the additional assumption (8.36), all of the terms in equations (8.20) will be $\mathcal{T}_{\Gamma_0}$-invariant and, consequently, the operator $\mathbb{L}$ will be also $\mathcal{T}_{\Gamma_0}$-invariant. Due to the uniqueness part of the Banach contraction principle, we conclude then that the fixed point of that map should be $\mathcal{T}_{\Gamma_0}$-invariant which, in turns, gives (8.37).

REMARK 8.12. Although we state the weighted estimate (8.27) for the weight $e^{-\gamma|x|}$ only, it holds also for all shifted weights $e^{-\gamma|x-x_0|}$ uniformly with respect to $x_0 \in \mathbb{R}^n$, see Remarks 4.11, 5.9 and 8.4. This, in turns, allows to prove (8.27) for all weights of exponential growth rates less or equal γ.

REMARK 8.13. It is worth noting that, since different Sobolev norms are equivalent on $\mathbb{P}(L)$, see Corollary 4.4, the manifold is, obviously, *precompact* in the local topology of $L^\infty_{e^{-\gamma|x|}}(\mathbb{R}^n)$. However, it is not *closed* in that topology since the pulses can escape to infinity and their number can decrease. If the number of non-escaping pulses remains infinite, the limit point will belong to $\mathbb{P}(L) = \mathbb{P}_\infty(L)$ or to its boundary $\partial\mathbb{P}_\infty(L)$, but if it remains only finite number of pulses, we obtain the finite-pulse manifold which is formally not contained in $\mathbb{P}_\infty(L)$

$$\mathbb{P}_N(L) := \{\mathbf{m} := \sum_{j=1}^{N} V_{\Gamma_j}, \ \ \operatorname{dist}(\xi_i, \xi_j) > 2L, \ \ i \neq j\}.$$

Thus the closure of $\mathbb{P}(L)$ in the local topology can be described as follows:

$$\bar{\mathbb{P}}(L) = \operatorname{cl}_{\mathrm{loc}}(\mathbb{P}(L)) = \mathbb{P}_\infty(L) \cup \partial\mathbb{P}_\infty(L) \cup \sum_{N=1}^{\infty} (\mathbb{P}_N(L) \cup \partial\mathbb{P}_N(L)) \cup \{0\} \tag{8.38}$$

(where zero corresponds to the case where all of the pulses escape to infinity).

Moreover, since the function $\mathbb{W}(t,\cdot)$ and all the terms in equations (8.20) are uniformly Lipschitz continuous in the weighted metric, they can be extended in a unique way from $\mathbb{P}(L)$ to $\bar{\mathbb{P}}(L)$ preserving all of the properties formulated in Theorem 8.5. In particular, restrictions of the extended $\mathbb{W}$ to the stratus $\mathbb{P}_N(L)$ give the center manifold reduction for the finite number N of pulses. For $N = 0$, we have $u_0(t) := \mathbb{W}(t, \{0\})$ which gives the perturbed trajectory (in the absence of pulses) associated with the hyperbolic equilibrium $u \equiv 0$ of the non-perturbed equation.

REMARK 8.14. The advantage the unified approach described in previous remark is, in particular, the possibility to use weighted estimates of the form (8.27) in situations where $\mathbf{m}_1$ and $\mathbf{m}_2$ contain *different* number of pulses. For instance, applying this estimate for $\mathbf{m}_1 = \mathbf{m}$ and $\mathbf{m}_2 = \{0\}$ where $\mathbf{m} \in \mathbb{P}(L)$ is some fixed pulse configuration, we will have

$$\|\mathbb{W}(t,\mathbf{m}) - u_0(t)\|_{L^\infty_{e^{-\gamma|x-x_0|}}(\mathbb{R}^n)} \leq C e^{+\delta L}(\mu + e^{-\alpha L})\|\mathbf{m}\|_{L^\infty_{e^{-\gamma|x-x_0|}}(\mathbb{R}^n)}$$

uniformly with respect to $x_0 \in \mathbb{R}^n$ and $\mathbf{m} \in \mathbb{P}(L)$. Multiplying this inequality by $e^{+\gamma \operatorname{dist}(x_0,\Xi)/2}$ (where $\Xi = \Xi(\mathbf{m}) = \{\xi_j\}_{j=1}^{\infty}$ are the pulse centers) and taking into account that this function is a weight function of exponential growth rate $\gamma/2$, we obtain

$$(8.39)\quad \|\mathbb{W}(t,\mathbf{m}) - u_0(t)\|_{L^\infty_{e^{+\gamma \operatorname{dist}(x,\Xi)/2}}(\mathbb{R}^n)} \leq \\ \leq C\,e^{\delta L}(\mu + e^{-\alpha L})\|\mathbf{m}\|_{L^\infty_{e^{+\gamma \operatorname{dist}(x,\Xi)/2}}(\mathbb{R}^n)} \leq C_1\,e^{\delta L}(\mu + e^{-\alpha L})$$

where C_1 is independent of t and $\mathbf{m} \in \mathbb{P}(L)$. The last formula shows that far away from the pulse centers, the function $\mathbb{W}$ is uniformly exponentially close to the perturbed equilibrium $u_0(t)$ in the absence of pulses.

We conclude this section by studying the behavior of the solutions of (8.1) in a small neighborhood of the above constructed center manifold. For simplicity, we restrict ourselves to consider only the most physically relevant case of spectrally stable pulses. See [**Mie90**] for cases where also unstable directions are present.

THEOREM 8.15. *Let the assumptions of Theorem 8.5 hold and let, in addition, the single pulse $V = V(x)$ be spectrally stable (see Section 2). Then, there exist $\delta > 0$ such that, for every initial data $(\mathbf{m}_\tau, v_\tau)$ such that*

$$(8.40)\qquad \mathbf{m}_\tau \in \mathbb{P}(L), \quad \|v_\tau\|_{\mathbb{X}_b} \leq \delta, \quad \mathbb{P}_{\mathbf{m}_\tau} v_\tau = 0$$

the associated semi-trajectory $(\mathbf{m}(t), v(t))$, $t \geq \tau$ of (8.20) exists globally in time and converges exponentially to the center manifold of this system. Moreover, there exists a trace-trajectory $\mathbf{m}_0(t)$ of the reduced system (8.26) such that

$$(8.41)\qquad \|v(t) - \mathbb{W}(t, \mathbf{m}_0(t))\|_{\mathbb{X}_b} + \|\mathbf{m}(t) - \mathbf{m}_0(t)\|_{L^\infty} \leq C\,e^{-\gamma(t-\tau)}, \quad t \geq \tau$$

where the constants C and γ are independent of $(\mathbf{m}_\tau, v_\tau)$, τ and t.

PROOF. The assertion of the theorem is a standard corollary of the normal hyperbolicity of the center manifolds. Nevertheless, since, in contrast to usual situation, the manifold is now *infinite-dimensional*, we prefer to give a complete proof of that fact.

We start with the global solvability of system (8.20) near the center manifold.

LEMMA 8.16. *Let the above assumptions hold. Then, for sufficiently small δ, the trajectory $(\mathbf{m}(t), v(t))$ of system (8.20) starting from the initial data satisfying (8.40) exists globally in time $t \geq \tau$ and remains close to the center manifold:*

$$(8.42)\qquad \|v(t)\|_{\mathbb{X}_b} \leq C\delta, \quad t \geq \tau$$

where C is independent of the concrete choice of the trajectory, satisfying (8.40).

PROOF. Indeed, the local existence of a solution in the phase space $\mathbb{X}_b$ for (8.20) is obvious, since, the equation for v is a parabolic semilinear PDE and the second one (for $\mathbf{m}$) is generated by a globally bounded smooth vector field on a globally bounded manifold $\mathbb{P}(L)$. So, we only need to verify *a priori* estimate of the form (8.42) for the (potentially unbounded) v-component. Then, by standard arguments, the local solution can be globally extended for all $t \geq \tau$.

In order to obtain such estimate, we invert the linear part of the v-equation using Proposition 7.9 (here we need the assumption that the single pulse is stable) and the first estimate of (8.23). Then, we have

$$\|v\|_{C([\tau,T],\mathbb{X}_b)} \leq C\|v_\tau\|_{\mathbb{X}_b} + C\kappa_0(e^{-\alpha L} + \mu + \|v\|^2_{C([\tau,T],\mathbb{X}_b)})$$

uniformly with respect to all $T \geq \tau$. It remains to note that, for sufficiently small κ_0, this inequality allows to obtain a priori upper bound for the norm of v of the form (8.42) if the norm of the initial data v_τ is small enough. Lemma 8.16 is proven. □

Our next task is to study the perturbations of the second equation of (8.20) by the exponentially decaying in time terms. For simplicity, we set from now on $\tau = 0$ and consider the perturbed version of the second equation of (8.20):

$$\mathbf{m}'(t) = \tilde{\mathrm{f}}(t, \mathbf{m}, v(t) - \tilde{v}(t)) \tag{8.43}$$

where the perturbation $\tilde{v}$ has finite norm in the space $C_{\mathrm{e}^{+\gamma t}}(\mathbb{R}_+, \mathbb{X}_{\mathrm{b}})$

$$\|\tilde{v}\|_{C_{\mathrm{e}^{+\gamma t}}(\mathbb{R}_+,\mathbb{X}_{\mathrm{b}})} \leq \nu \tag{8.44}$$

and $\gamma > 0$ is large enough.

LEMMA 8.17. *Let the function v satisfy inequality* (8.42) *for sufficiently small $\delta > 0$. Let us assume also that the quantity $\gamma_0 := C_\varepsilon(\mathrm{e}^{-2(\alpha-\varepsilon)L} + \mu + \delta + \nu)$ is small enough. Then, for every $\gamma > \gamma_0$ and every function $\tilde{v} \in C_{\mathrm{e}^{+\gamma t}}(\mathbb{R}_+, \mathbb{X}_{\mathrm{b}})$ satisfying* (8.44), *there exists a unique solution $\mathbf{m}_0 = \mathbf{m}_0(\tilde{v})$ of problem* (8.43) *such that*

$$\|\mathbf{m} - \mathbf{m}_0\|_{C_{\mathrm{e}^{+\gamma t}}(\mathbb{R}_+,\mathbb{X}_{\mathrm{b}})} \leq \frac{\gamma_0}{\gamma - \gamma_0}\|\tilde{v}\|_{C_{\mathrm{e}^{+\gamma t}}(\mathbb{R}_+,\mathbb{X}_{\mathrm{b}})}. \tag{8.45}$$

Moreover, the map $\tilde{v} \to \mathbf{m}_0(\tilde{v})$ is uniformly Lipschitz continuous in the weighted metrics:

$$\|\mathbf{m}_0(\tilde{v}_1) - \mathbf{m}_0(\tilde{v}_2)\|_{C_{\mathrm{e}^{+\gamma t}}(\mathbb{R}_+,\mathbb{X}_{\mathrm{b}})} \leq \frac{\gamma_0}{\gamma - \gamma_0}\|\tilde{v}_1 - \tilde{v}_2\|_{C_{\mathrm{e}^{+\gamma t}}(\mathbb{R}_+,\mathbb{X}_{\mathrm{b}})} \tag{8.46}$$

for all $\tilde{v}_1$ and $\tilde{v}_2$ satisfying (8.44).

PROOF. We first note that, due to Lemma 8.2 and Theorem 6.6, we have

$$\|\tilde{\mathrm{f}}(t, \mathbf{m}_1, v(t) - \tilde{v}(t)) - \tilde{\mathrm{f}}(t, \mathbf{m}_2, v(t))\|_{L^\infty(\mathbb{R}^n)} \leq \gamma_0(\|\tilde{v}(t)\|_{\mathbb{X}_{\mathrm{b}}} + \|\mathbf{m}_1 - \mathbf{m}_2\|_{L^\infty(\mathbb{R}^n)}) \tag{8.47}$$

uniformly with respect to $\mathbf{m}_i \in \mathbb{P}(L)$. For every $N \in \mathbb{N}$, we now define the function $\mathbf{m}_N(t)$ as a unique backward solution of the following problem

$$\mathbf{m}_N'(t) = \tilde{\mathrm{f}}(t, \mathbf{m}_N(t), v(t) - \tilde{v}(t)), \quad \mathbf{m}_N(N) = \mathbf{m}(N), \quad t \in [0, N].$$

Then, due to estimate (6.4), we have

$$\|\mathbf{m}_N(t) - \mathbf{m}(t)\|_{L^\infty(\mathbb{R}^n)} \leq \gamma_0 \int_t^N \mathrm{e}^{\gamma_0(s-t)}\,\|\tilde{v}(s)\|_{\mathbb{X}_{\mathrm{b}}}\,ds, \quad t \in [0, N].$$

Extending $\mathbf{m}_N(t)$ for $t \geq N$ by $\mathbf{m}_N(t) = \mathbf{m}(t)$, multiplying the last inequality by $\mathrm{e}^{+\gamma t}$ and taking sup over $t \in \mathbb{R}_+$, we have

$$\|\mathbf{m}_N - \mathbf{m}\|_{C_{\mathrm{e}^{+\gamma t}}(\mathbb{R}_+,\mathbb{X}_{\mathrm{b}})} \leq \frac{\gamma_0}{\gamma - \gamma_0}\|\tilde{v}\|_{C_{\mathrm{e}^{+\gamma t}}(\mathbb{R}_+,\mathbb{X}_{\mathrm{b}})} \tag{8.48}$$

and, analogously, for different $\tilde{v}$, we have

$$\|\mathbf{m}_N(\tilde{v}_1) - \mathbf{m}_N(\tilde{v}_2)\|_{C_{\mathrm{e}^{+\gamma t}}(\mathbb{R}_+,\mathbb{X}_{\mathrm{b}})} \leq \frac{\gamma_0}{\gamma - \gamma_0}\|\tilde{v}_1 - \tilde{v}_2\|_{C_{\mathrm{e}^{+\gamma t}}(\mathbb{R}_+,\mathbb{X}_{\mathrm{b}})}. \tag{8.49}$$

We claim that $\{\mathbf{m}_N\}_{N\in\mathbb{N}}$ is a Cauchy sequence $C_{\mathrm{e}^{(\gamma-\varepsilon)t}}(\mathbb{R}_+, L^\infty(\mathbb{R}^n))$ as $N \to \infty$. Indeed, for $M \geq N$, according to (8.49), we have

$$\|\mathbf{m}_N - \mathbf{m}_M\|_{C_{\mathrm{e}^{(\gamma-\varepsilon)t}}(\mathbb{R}_+,L^\infty(\mathbb{R}^n))} \leq \frac{\gamma_0}{\gamma - \varepsilon - \gamma_0} \sup_{t\in[N,M]} \{\mathrm{e}^{(\gamma-\varepsilon)t}\,\|\tilde{v}\|_{\mathbb{X}_{\mathrm{b}}}\} \leq C\,\mathrm{e}^{-\varepsilon N}.$$

Thus, we can define the limit solution $\mathbf{m}_0(t)$ of problem (8.43) by $\mathbf{m}_0=\lim_{N\to\infty}\mathbf{m}_N$ (where the limit is taken e.g. in $C_b(\mathbb{R}_+,L^\infty(\mathbb{R}^n))$. Since estimates (8.48) and (8.49) are uniform with respect to N, then the limit function $\mathbf{m}_0$ also satisfies these estimates. Thus, the required solution $\tilde v\to\mathbf{m}_0(\tilde v)$ satisfying (8.46) and (8.47) is constructed. The uniqueness of that solution is obvious since the decaying exponent γ is assumed to be larger than the Lipschitz constant γ_0 of equation (8.43). Lemma (8.17) is proven. □

Let us now construct the v-component associated with the $\mathbf{m}$-component $\mathbf{m}_0(t)$ constructed in the previous lemma. We seek for it as a solution of the following equation:

$$\partial_t v_0+A_0v_0+\Phi'(\mathbf{m}_0(t))v_0=\mathbb{H}(t,\mathbf{m}_0(t),v(t)-\tilde v(t)),\quad v_0(0)=\mathbb{W}(0,\mathbf{m}_0(0)) \tag{8.50}$$

where $\mathbf{m}_0=\mathbf{m}_0(\tilde v)$ is defined in the previous lemma. The next lemma gives the natural estimates for the solution $v_0(t)$ thus defined.

LEMMA 8.18. *Let the above assumptions hold. Then, for sufficiently small μ, ν and δ and sufficiently large L, the following estimate holds:*

$$\|v-v_0\|_{C_{e^{+\gamma t}}(\mathbb{R}_+,\mathbb{X}_b)}\le C\delta+C_1\gamma_0(1+\gamma_0)\nu \tag{8.51}$$

where $\gamma_0:=C_\varepsilon(e^{-2(\alpha-\varepsilon)L}+\mu+\nu+\delta)$, $\gamma>2\gamma_0$ is some positive exponent and C and C_1 are some constants independent of δ, ν and γ_0.

Moreover, the map $\tilde v\to v_0$ is uniformly Lipschitz continuous in the following sense:

$$\|v_0(\tilde v_1)-v_0(\tilde v_2)\|_{C_{e^{+\gamma t}}(\mathbb{R}_+,\mathbb{X}_b)}\le C_1\gamma_0(1+\gamma_0)\|\tilde v_1-\tilde v_2\|_{C_{e^{+\gamma t}}(\mathbb{R}_+,\mathbb{X}_b)} \tag{8.52}$$

for all $\tilde v_1$ and $\tilde v_2$ satisfying the above properties.

PROOF. Indeed, due to Theorem 8.5 and Lemma 8.17, we have

$$\|v_0(0)\|_{\mathbb{X}_b}\le C\kappa_0\gamma_0\nu$$

and the analogous estimate for $\|v_0(\tilde v_1)(0)-v_0(\tilde v_2)(0)\|_{\mathbb{X}_b}$. Proposition 7.3 and Lemma 8.3 then give

$$\|v_0\|_{C_b(\mathbb{R}_+,\mathbb{X}_b)}\le C(\delta+\nu) \tag{8.53}$$

if γ_0 and κ_0 is small enough. This estimate, together with Lemma 8.3 yield

$$\begin{aligned}\|\mathbb{H}(t,\mathbf{m}_0(t),v(t)-\tilde v(t))-\mathbb{H}(t,\mathbf{m}(t),v(t))\|_{L^p_b(\mathbb{R}^n)}\le\\ \le C\gamma_0(\|\tilde v(t)\|_{\mathbb{X}_b}+\|\mathbf{m}(t)-\mathbf{m}_0(t)\|_{\mathbb{X}_b})\end{aligned} \tag{8.54}$$

and, analogously, for the Lipschitz continuity

$$\begin{aligned}\|\mathbb{H}(t,\mathbf{m}_0(\tilde v_1)(t),v(t)-\tilde v_1(t))-\mathbb{H}(t,\mathbf{m}_0(\tilde v_2)(t),v(t)-\tilde v_2(t))\|_{L^p_b(\mathbb{R}^n)}\le\\ \le C\gamma_0(\|\tilde v_1(t)-\tilde v_2(t)\|_{\mathbb{X}_b}+\|\mathbf{m}_0(\tilde v_1)(t)-\mathbf{m}_0(\tilde v_2)(t)\|_{\mathbb{X}_b}).\end{aligned} \tag{8.55}$$

Then, for the difference $\bar v(t)=v(t)-v_0(t)$, we have

$$\begin{aligned}\partial_t\bar v+A_0\bar v+\Phi'(\mathbf{m}(t))\bar v=[\Phi'(\mathbf{m}(t))-\Phi'(\mathbf{m}_0(t))]v_0(t)+\\ +[\mathbb{P}_{\mathbf{m}}-\mathbb{P}_{\mathbf{m}_0}]v_0(t)+[\mathbb{H}(t,\mathbf{m}(t),v(t))-\mathbb{H}(t,\mathbf{m}_0(t),v(t)-\tilde v(t))],\quad \bar v(t)=v(0)-v_0(0).\end{aligned}$$

Applying Proposition 7.3 to this equation, using (8.55) and (8.53) and fixing γ in such way that the weight $e^{\gamma t}$ satisfies the assumptions of that proposition, we establish estimate (8.51) (here we have also implicitly used that $\|\mathbf{m}_1-\mathbf{m}_2\|_{\mathbb{X}_b}\le$

$C\|\mathbf{m}_1-\mathbf{m}_2\|_{L^\infty(\mathbb{R}^n)})$. The Lipschitz continuity (8.52) can be established completely analogous. Lemma 8.18 is proven. □

We are now ready to finish the proof of the theorem. Indeed, let us consider the ν-ball $\mathcal{B}_\nu$ in the space $C_{e^{\gamma t}}(\mathbb{R}_+,\mathbb{X}_b)$ and the map

$$\mathcal{Q}:\mathcal{B}_\nu\to C_{e^{\gamma t}}(\mathbb{R}_+,\mathbb{X}_b),\quad \mathcal{Q}(\tilde v):=v-v_0(\tilde v).$$

We claim that, for sufficiently small δ and μ and large L, there exists ν_0 such that $\mathcal{Q}:\mathcal{B}_{\nu_0}\to\mathcal{B}_{\nu_0}$. Indeed, due to Lemma 8.18, it is sufficient to take $\nu_0=2C\delta$ if δ, L and μ are such that $C_1\gamma_0(1+\gamma_0)<1/2$. Moreover, estimate (8.52) shows that, in this case, the map $\mathcal{Q}:\mathcal{B}_{\nu_0}\to\mathcal{B}_{\nu_0}$ will be a contraction.

Thus, due to Banach contraction principle, there exist a (unique) fixed point $\tilde v_0$ of this map. This means that the function $v_0(\tilde v_0)=v-\tilde v_0$ satisfies

$$\partial_t v_0+Av_0+\Phi'(\mathbf{m}_0)v_0+\mathbb{P}_{\mathbf{m}_0}v_0=\mathbb{H}(t,\mathbf{m}_0(t),v_0(t))$$

where $\mathbf{m}_0=\mathbf{m}_0(\tilde v_0)$ and, therefore, the pair $(\mathbf{m}_0,v_0)$ solves indeed problem (8.20). Moreover, since $v_0(0)=\mathbb{W}(0,\mathbf{m}_0(0))$, then

$$v_0(t)=\mathbb{W}(t,\mathbf{m}_0(t))),\quad \forall t\ge 0.$$

Thus, $(\mathbf{m}_0,v_0)$ belongs to the center manifold. Finally, estimate (8.41) is now an immediate corollary of (8.45) and (8.51) and Theorem 8.15 is proven. □

REMARK 8.19. The assertion of the last theorem can be easily reformulated in terms of the initial problem (8.1). Namely, under the assumptions of Theorem 8.15, there exist a constant δ such that, if $u_\tau\in\mathcal{O}_\delta(\mathbb{P}(L'))$ then either the associated solution $u(t)$ of problem (8.1) belongs to $\mathcal{O}_{C\delta}(\mathbb{P}(L'))$ for all $t\ge\tau$ or there exists $T_*\ge\tau$ such that $u(T_*)\in\mathcal{O}_{C\delta}(\partial\mathbb{P}(L'))$ (in the first case we set $T_*=\infty$).

In both cases, there exists a trajectory of (8.1) on the center manifold $\widetilde u(t)\in\mathbb{U}_{L'}(t)$ for $t\in(\tau,T_*)$ such that

$$\|u(t)-\widetilde u(t)\|_{\mathbb{X}_b}\le C\,e^{-\gamma(t-\tau)},\quad t\in[\tau,T_*)$$

where C and $\gamma>0$ are independent of u_τ, τ and t. Thus, in the spectrally stable case, the dynamics generated by (8.1) in the small neighborhood of the multi-pulse manifold $\mathbb{P}(L')$ is completely determined by the reduced dynamics on the center manifold generated by (8.26).

9. Hyperbolicity and stability

The main task of this section is to discuss briefly the basic facts of hyperbolic theory adapted to the case of infinite-dimensional equations of the form (8.1) or (8.26), see e.g. [**KaH95, MiZ04**] for more details. We start with equation (8.1) which we rewrite in the following form:

$$\partial_t u+A_0u+F(t,u)=0 \tag{9.1}$$

with $F(t,u):=\Phi(u)-\mu R(t,u)$ and $R(t,u)=R(t,x,u,D_xu,\cdots,D_x^{2l-1}u)$ satisfying the assumptions of Section 8.

DEFINITION 9.1. A complete trajectory $u\in W_b^{(1,2l),p}(\mathbb{R}^{n+1})$ of equation (9.1) is hyperbolic if there exists a constant $C_u>0$ such that, for every $h\in L^p_b(\mathbb{R}^{n+1})$, the associated inhomogeneous equation of variations

$$\partial_t w+A_0w+F'_u(t,u(t))w=h \tag{9.2}$$

has a unique solution $w \in W_{\mathrm{b}}^{(1,2l),p}(\mathbb{R}^{n+1})$ and the following estimate holds:

$$\|w\|_{W_{\mathrm{b}}^{(1,2l),p}(\mathbb{R}^{n+1})} \le C_u \|h\|_{L_{\mathrm{b}}^p(\mathbb{R}^{n+1})}. \tag{9.3}$$

A set $\mathcal{H}^{tr} \subset W_{\mathrm{b}}^{(1,2l),p}(\mathbb{R}^{n+1})$ of trajectories of (9.1) is a (uniformly) hyperbolic (trajectory) set if it is bounded

$$\|\mathcal{H}^{tr}\|_{W_{\mathrm{b}}^{(1,2l),p}(\mathbb{R}^{n+1})} \le C^1_{\mathcal{H}^{tr}} \tag{9.4}$$

and every trajectory $u \in \mathcal{H}^{tr}$ is hyperbolic and the hyperbolicity constant C_u in (9.3) is independent of the concrete choice of $u \in \mathcal{H}^{tr}$ ($C_u \le C_{\mathcal{H}^{tr}}$).

REMARK 9.2. The above definition of a hyperbolic trajectory is equivalent to the standard one via stable and unstable foliations. In particular, it is not difficult to verify that this definition is independent of the concrete choice of the exponent p. We however emphasize once more that this definition has sense for the *nonautonomous and non-homogeneous* (e.g. space-time periodic) case only. In contrast to this, in the autonomous or homogeneous case, the functions $\partial_t u$ or/and $\nabla_x u$ always belong to the kernel of equation of variations, so the *neutral* foliation appears. Furthermore, the standard requirement for the finite-dimensional theory is that this foliation is one-dimensional (or, more generally, finite-dimensional). This can be satisfied only for spatially localized structures decaying exponentially as $x \to \infty$. Obviously, this requirement is too restrictive for the study the multi-pulse structures containing infinitely many pulses. Moreover, we do not know any reasonable extention of the hyperbolic theory for that case. That is the reason why we restrict ourselves to consider only the space-time non-homogeneous case where such a theory exists and is analogous to the finite-dimensional case.

The next theorem gives the standard relation between hyperbolicity and stability of a hyperbolic trajectory.

THEOREM 9.3. *Let equation* (9.1) *satisfy the above assumptions. Then a complete trajectory* $u \in W_{\mathrm{b}}^{(1,2l),p}(\mathbb{R}^{n+1})$ *is hyperbolic in the sense of previous definition if and only if, there exists, a neighborhood* $\mathcal{V}_\delta(u)$ *of* u *in* $W_{\mathrm{b}}^{(1,2l),p}(\mathbb{R}^{n+1})$ *and for every function* $\tilde F(t,u) := \tilde F(t,x,u,D_x u,\cdots,D^{2l-1}u)$ *which belongs to* L^p *with respect to* (t,x) *and* C^1 *with respect to* u, *such that the norm*

$$\|\tilde F\| := \|\tilde F(\cdot,\cdot)\|_{C^1(V_\delta(u),L^p(\mathbb{R}^{n+1}))} < \delta_u \tag{9.5}$$

is small enough, the perturbed equation

$$\partial_t u + A_0 + F(t,u) = \tilde F(t,u) \tag{9.6}$$

has a unique solution $\tilde u$ *in* $V_\delta(u)$ *and this solution satisfies*

$$\|u - \tilde u\|_{W_{\mathrm{b}}^{(1,2l),p}(\mathbb{R}^{n+1})} \le C\|\tilde F\| \tag{9.7}$$

with the constant C *independent of* $\tilde F$.

Moreover, the perturbed trajectory $\tilde u$ *is also hyperbolic and the set* $\mathcal{H}^{tr}$ *is uniformly hyperbolic if and only if it is bounded and the constant* C *in* (9.7) *is independent of the concrete choice of* $u \in \mathcal{H}^{tr}$.

PROOF. Indeed, let the trajectory u be hyperbolic in the sense of Definition 9.1. We will seek for the desired solution of the perturbed equation (9.6) in the form $\tilde{u} := u + v$. Then, the function v should satisfy the following equation:

$$\text{(9.8)}\quad \partial_t v + A_0 v + F'_u(t, u(t))v = -[F(t, u + v) - F(t, u) - F'(t, u)v] + \tilde{F}(t, u + v).$$

Since the trajectory u is bounded in $W_{\mathrm{b}}^{(1,2l),p}(\mathbb{R}^{n+1})$ and F is smooth, the function $H(t, v) := F(t, u + v) - F(t, u) - F'(t, u)v$ satisfies

$$\|H(\cdot, v)\|_{L^p(\mathbb{R}^{n+1})} + \|H'_v(\cdot, v)\|_{\mathcal{L}(W_{\mathrm{b}}^{(1,2l),p}(\mathbb{R}^{n+1}),L^p(\mathbb{R}^{n+1}))} \le C_1\|v\|_{W_{\mathrm{b}}^{(1,2l),p}(\mathbb{R}^{n+1})}$$

where C_1 depends only on F and the norm of u.

Using now that the linear part of equation (9.8) is invertible (the hyperbolicity assumption) and applying the implicit function theorem, we deduce the existence and uniqueness of the required solution $\tilde{u}$ and estimate (9.7). Note also that the constant C in (9.7) depends only on the hyperbolicity constant C_u, function F and the norm of u and is independent of the concrete choice of u and $\tilde{F}$.

Let us now assume that the trajectory $u \in W_{\mathrm{b}}^{(1,2l),p}(\mathbb{R}^{n+1})$ is stable in the sense that (9.7) holds for every sufficiently small perturbation $\tilde{F}$ and the associated perturbed solution $\tilde{u}$. We need to verify that equation (9.2) is uniquely solvable for every $h \in L^p_{\mathrm{b}}(\mathbb{R}^{n+1})$ and estimate (9.3) holds. To this end, for every sufficiently small $\varepsilon > 0$, we consider the perturbation $\tilde{F} = \varepsilon h(t)$. Let u_ε be the associated solution (9.6) and $w_\varepsilon := \frac{u_\varepsilon - u}{\varepsilon}$. Then, this function satisfies

$$\text{(9.9)}\quad \partial_t w_\varepsilon + A w_\varepsilon + F'_u(t, u(t))w_\varepsilon - h = -\frac{1}{\varepsilon}[F(t, u + \varepsilon w_\varepsilon) - F(t, u) - \varepsilon F'_u(t, u)w_\varepsilon]$$

and, due to estimate (9.7), we have

$$\text{(9.10)}\qquad \|w_\varepsilon\|_{W_{\mathrm{b}}^{(1,2l),p}(\mathbb{R}^{n+1})} \le C\|h\|_{L^p(\mathbb{R}^{n+1})}.$$

We see that the right-hand side of equation (9.9) tends to zero as $\varepsilon \to 0$ (in the L^p_{b}-norm). Moreover, due to (9.10), we may assume without loss of generality that $w_\varepsilon \to w$ weakly in $W_{\mathrm{loc}}^{(1,2l),p}(\mathbb{R}^{n+1})$. Passing then to the limit in equation (9.9), we conclude that the function w solves (9.2) and (9.10) gives estimate (9.3) for this solution. Thus, the existence of a solution of (9.2) and estimate (9.3) are verified.

Let us now verify the uniqueness. Indeed, let the uniqueness fail. Then there exists $w_0 \in W_{\mathrm{b}}^{(1,2l),p}(\mathbb{R}^{n+1})$ such that

$$\text{(9.11)}\qquad \partial_t w_0 + A_0 w_0 + F'_u(t, u)w_0 = 0.$$

Let us now construct the perturbed equation (9.6) in order the function $\tilde{u} = u + \varepsilon w_0$ to be its solution. Indeed, it is not difficult to verify using (9.11) that $\tilde{u}$ solves

$$\partial_t \tilde{u} + A_0 \tilde{u} + F(t, \tilde{u}) = \tilde{F}(t, \tilde{u}) := [F(t, u + \varepsilon w_0) - F(t, u) - \varepsilon F'(t, u)w_0] := \varepsilon^2 h_\varepsilon(t).$$

Obviously, $\|h_\varepsilon\|_{L^p_{\mathrm{b}}(\mathbb{R}^{n+1})} \le C$ and, consequently, due to (9.7),

$$\|\tilde{u} - u\|_{W_{\mathrm{b}}^{(1,2l),p}(\mathbb{R}^{n+1})} = \varepsilon\|w_0\|_{W_{\mathrm{b}}^{(1,2l),p}(\mathbb{R}^{n+1})} \le C_1\varepsilon^2.$$

Passing to the limit $\varepsilon \to 0$ in this formula, we see that $\|w_0\|_{W_{\mathrm{b}}^{(1,2l),p}(\mathbb{R}^{n+1})} = 0$ and the uniqueness is also verified. Theorem 9.3 is proven. $\square$

REMARK 9.4. It follows from the proof of Theorem 9.3 that, in order to verify the hyperbolicity, it is sufficient to prove the existence of $\tilde{u}$ and estimate (9.7) only for the perturbations of the form

$$\tilde{F}(t,u) = \varepsilon h(t), \quad h \in L^p_{\rm b}(\mathbb{R}^{n+1}), \quad \varepsilon \ll 1.$$

Moreover, it is sufficient to consider only more regular than $L^p_{\rm b}$-integrable functions h, say, belonging to $C_{\rm b}(\mathbb{R}^{n+1})$ or $C_{\rm b}(\mathbb{R}, \mathbb{X}_{\rm b}(\mathbb{R}^n))$. Indeed, in this case, arguing as before, instead of (9.2), we will have an estimate

$$\|w\|_{W_b^{(1,2l),p}(\mathbb{R}^{n+1})} \le C\|h\|_{C_{\rm b}(R, C^{2l-1}_{\rm b}(\mathbb{R}^n))} \tag{9.12}$$

and it remains to note that, due to the parabolic regularity, the solvability of (9.1) for more regular h and estimate (9.12) imply the solvability for $h \in L^p_{\rm b}(\mathbb{R}^{n+1})$ and estimate (9.3). In order to see that, one can use that, for sufficiently large constant M, the auxiliary equation

$$\partial_t w_1 + A_0 w_1 + F'(t, u(t))w_1 + M w_1 = h \tag{9.13}$$

for all $h \in L^p_{\rm b}(\mathbb{R}^{n+1})$ and satisfies (9.3) and the remainder $w_2 = w - w_1$ satisfies the analog of (9.2)

$$\partial_t w_2 + A_0 w_2 + F'_u(t, u(t))w_2 = M w_1$$

with more regular external forces $M w_1 \in C_{\rm b}(\mathbb{R}, C^{2l-1}_{\rm b}(\mathbb{R}^n))$ for which (9.12) is applicable.

Moreover, due to the smoothing property for parabolic equations, we may replace the $W_b^{(1,2l),p}(\mathbb{R}^{n+1})$-norm in the left-hand side of (9.7) by a weaker norm, for instance $C_{\rm b}(\mathbb{R}, C^{2l-1}_{\rm b}(\mathbb{R}^n))$. We will use these simple observations below in order to verify that a hyperbolic set for the reduced system on the pulse manifold $\mathbb{P}(L)$ remains hyperbolic for the whole system (8.1) as well.

Thus, if $\mathcal{H}^{tr}$ a hyperbolic set of equation (9.1) and $\tilde{F}$ is a sufficiently small perturbation, then, according to Theorem 9.3, for every $u \in \mathcal{H}^{tr}$ there exist a unique hyperbolic trajectory $\tilde{u} \in \tilde{\mathcal{H}}^{tr}$ of the perturbed equation and, therefore the map

$$\mathcal{S} : \mathcal{H}^{tr} \to \tilde{\mathcal{H}}^{tr}, \quad \mathcal{S}u = \tilde{u} \tag{9.14}$$

is well defined. The next corollary shows that $\mathcal{S}$ is a homeomorphism in the local topology.

COROLLARY 9.5. *Let $\mathcal{H}^{tr}$ be a hyperbolic set of equation (9.1) and let the perturbation $\tilde{F}$ be sufficiently small. Then the map $\mathcal{S} : \mathcal{H}^{tr} \to \tilde{\mathcal{H}}^{tr}$ defined via (9.14) is a homeomorphism in the topology of $W^{(1,2l),p}_{\rm loc}(\mathbb{R}^{n+1})$.*

PROOF. We first note that the sets $\mathcal{H}^{tr}$ and $\tilde{\mathcal{H}}^{tr}$ are precompact in the space $W^{(1,2l),p}_{\rm loc}(\mathbb{R}^{n+1})$. Indeed, since $\mathcal{H}^{tr}$ is bounded in $W_b^{(1,2l),p}(\mathbb{R}^{n+1})$, it is precompact in $C_{\rm loc}(\mathbb{R}, C^{2l-1}(\mathbb{R}^n))$ (due to our choice of the exponent p). Using now the fact that the map $u \to F(\cdot, u)$ is continuous as the map from $C_{\rm loc}(\mathbb{R}, C^{2l-1}(\mathbb{R}^n))$ to $L^p_{\rm loc}(\mathbb{R}^{n+1})$ together with the parabolic regularity (3.9), we conclude that $\mathcal{H}^{tr}$ is compact in $W^{(1,2l),p}_{\rm loc}(\mathbb{R}^{n+1})$. The (pre)compactness of $\tilde{\mathcal{H}}^{tr}$ can be established analogously. Moreover, without loss of generality, we can assume that they are closed in $W^{(1,2l),p}_{\rm loc}(\mathbb{R}^{n+1})$, otherwise we can take a closure preserving the uniform hyperbolicity. Thus, $\mathcal{H}^{tr}$ and $\tilde{\mathcal{H}}^{tr}$ are compact.

Thus, we only need to verify the continuity of $\mathcal{S}$. Indeed, let $\{u_n\}_{n=1}^\infty \subset \mathcal{H}^{tr}$ be a sequence converging to some $u \in \mathcal{H}^{tr}$ in $W^{(1,2l),p}_{\rm loc}(\mathbb{R}^{n+1})$ and let $\tilde u_n = \mathcal{S}u_n \in \tilde{\mathcal{H}}^{tr}$. Let also $\tilde u$ belongs to the limit set of $\{\tilde u_n\}$ (it exists since $\tilde{\mathcal{H}}^{tr}$ is compact). Then, due to estimate (9.7), we have

$$\|\tilde u_n - u_n\|_{W^{(1,2l),p}_{\rm loc}(\mathbb{R}^{n+1})} \le C\|\tilde F\|$$

and, consequently,

$$\|\tilde u - u\|_{W^{(1,2l),p}_{\rm loc}(\mathbb{R}^{n+1})} \le C\|\tilde F\|.$$

Since $\mathcal{S}u$ is a *unique* solution of the perturbed equation belonging to $V_\delta(u)$, then necessarily $\tilde u = \mathcal{S}u$ (if the perturbation is small enough). Thus, $\{\tilde u_n\}_{n=1}^\infty$ has a unique limit point $\tilde u = \lim_{n\to\infty} \tilde u_n = \mathcal{S}u$. Corollary 9.5 is proven. □

REMARK 9.6. The result of Corollary 9.5 can be improved as follows: there exists positive ε_0 such that, for every weight function $\theta(t,x)$ of exponential growth rate $\varepsilon \le \varepsilon_0$, the following estimate holds:

$$C^{-1}\|u_1-u_2\|_{W^{(1,2l),p}_\theta(\mathbb{R}^{n+1})} \le \|\mathcal{S}u_1-\mathcal{S}u_2\|_{W^{(1,2l),p}_\theta(\mathbb{R}^{n+1})} \le C\|u_1-u_2\|_{W^{(1,2l),p}_\theta(\mathbb{R}^{n+1})},$$

see [**MiZ04**]. However, the proof of this estimate given there essentially uses that the phase space is a linear space. So, keeping in mind the extention of the hyperbolic theory to equations on the pulse *manifold* $\mathbb{P}(L)$, we prefer to give the alternative proof of Corollary 9.5 which does not use the global linear structure and can be immediately extended to manifolds.

REMARK 9.7. Projecting the trajectory hyperbolic set $\mathcal{H}^{tr}$ to the space $\mathbb{X}_{\rm b} := W^{2l(1-1/p),p}_b(\mathbb{R}^n)$ via $Tr\big|_{t=\tau}u = u(\tau)$, we obtain the associated hyperbolic sets $\mathcal{H}_\tau := Tr\big|_{t=\tau}\mathcal{H}^{tr}$ in the phase space for every $\tau \in \mathbb{R}$. If it is known, in addition, that the map $Tr\big|_{t=\tau}$ is injective, then, due to compactness it will be a homeomorphism of sets $\mathcal{H}^{tr}$ and $\mathcal{H}_\tau$ in the local topology and, consequently, due to Corollary 9.5 the sets $\mathcal{H}_\tau$ and $\tilde{\mathcal{H}}_\tau$ will be also homeomorphic. The sufficient conditions for the injectivity of $Tr\big|_{t=\tau}$ (=backward uniqueness) can be found e.g. in [**AgN63**] and [**AgN67**] although we do not know whether or not the backward uniqueness theorem holds for solutions of (9.1) under the above general assumptions. We however note that, for the hyperbolic sets obtained from the center manifold reduction to the multi-pulse manifolds (which we are mainly interested in this paper) such uniqueness is immediate since it takes place on the center manifold. So, we need not to verify rather delicate backward uniqueness property for that particular case.

Our next task is to extend the above theory to equations on the multi-pulse manifold $\mathbb{P}(L)$ analogous to (8.26). To be more precise, we consider the following equation on $\mathbb{P}(L)$:

$$\frac{d}{dt}\mathbf{m} = \mathrm{f}(t,\mathbf{m}), \quad \mathbf{m} \in \mathbb{P}(L) \tag{9.15}$$

with the uniformly smooth and bounded function f.

DEFINITION 9.8. Analogously, we say that a complete trajectory $\mathbf{m}$ of equation (9.15) is hyperbolic if, for every $H \in L^\infty(\mathbb{R}^{n+1})$ such that $\mathbb{P}_{\mathbf{m}(t)}H(t) \equiv H(t)$, the associated equation of variation

$$\mathbb{P}_{\mathbf{m}(t)}\frac{d}{dt}\,\mathrm{w} - \mathbb{P}_{\mathbf{m}(t)}(\mathrm{f}'_{\mathbf{m}}(t,\mathbf{m}(t))\,\mathrm{w}) = H(t) \tag{9.16}$$

has a unique solution $\mathrm{w} \in L^\infty(\mathbb{R}, T_{\mathbf{m}(t)}\mathbb{P}(L))$ and the following estimate holds:

$$\|\,\mathrm{w}\,\|_{L^\infty(\mathbb{R}^{n+1}))} \leq C_{\mathbf{m}}\|H\|_{L^\infty(\mathbb{R}^{n+1})}. \tag{9.17}$$

A set $\mathcal{H}^{tr}$ of trajectories of equation (9.15) is a (uniformly) hyperbolic set if every trajectory belonging to it is hyperbolic and estimate (9.17) holds uniformly with respect to all trajectories from $\mathcal{H}^{tr}$ (we do not require boundedness of $\mathcal{H}^{tr}$ here since the manifold $\mathbb{P}(L)$ is globally bounded).

The following theorem is the analog of Theorem 9.3 for (9.15) on $\mathbb{P}(L)$.

THEOREM 9.9. *A complete trajectory* $\mathbf{m} \in L^\infty(\mathbb{R}^{n+1})$ *of equation* (9.15) *is hyperbolic in the sense of previous definition if and only if, there exists, a neighborhood* $\mathcal{V}_\delta(\mathbf{m})$ *of* u *in* $L^\infty(\mathbb{R}^{n+1})$ *and for every function* $\tilde{\mathrm{f}}(t,\mathbf{m})$ *such that* $\mathbb{P}_{\mathbf{m}}\tilde{\mathrm{f}}(t,\mathbf{m}) = \tilde{\mathrm{f}}(t,\mathbf{m})$ *which belongs to* L^∞ *with respect to* t *and* C^1 *with respect to* $\mathbf{m}$, *such that the norm*

$$\|\tilde{\mathrm{f}}\| := \sup_{t\in\mathbb{R}}\ \sup_{v\in V_\delta(\mathbf{m})} \left(\|\tilde{\mathrm{f}}(t,v)\|_{L^\infty(\mathbb{R}^n)} + \|\tilde{\mathrm{f}}'_v(t,v)\|_{\mathcal{L}(L^\infty(\mathbb{R}^n),L^\infty(\mathbb{R}^n))}\right) < \delta \tag{9.18}$$

is small enough, the perturbed equation

$$\frac{d}{dt}\mathbf{m} - \mathrm{f}(t,\mathbf{m}) = \tilde{\mathrm{f}}(t,\mathbf{m}) \tag{9.19}$$

has a unique solution $\tilde{\mathbf{m}}$ *in* $V_\delta(\mathbf{m})$ *and this solution satisfies*

$$\|\mathbf{m} - \tilde{\mathbf{m}}\|_{L^\infty(\mathbb{R}^{n+1})} \leq C\|\tilde{\mathrm{f}}\| \tag{9.20}$$

with the constant C *independent of* $\tilde{\mathrm{f}}$.

Moreover, the perturbed trajectory $\tilde{\mathbf{m}}$ *is also hyperbolic.*

PROOF. The proof of this result is very simular to the proof of Theorem 9.3. The only difference with the case of linear phase space is the fact that now the difference $\tilde{\mathbf{m}}(t) - \mathbf{m}(t)$ between the solutions of (9.15) and (9.19) does not belong to the tangent space and, therefore, cannot be directly interpreted as a solution of some equation on the tangent space close to the equation of variations. That is why, we only explain below how to overcome this problem leaving the details to the reader.

To be more precise, instead of the difference $\tilde{\mathbf{m}} - \mathbf{m}$ we will write the equation on its projection w to the tangent space

$$\mathrm{w}(t) := \mathbb{P}_{\mathbf{m}(t)}(\tilde{\mathbf{m}}(t) - \mathbf{m}(t)).$$

Then, on the one hand, arguing as in Theorem 7.1, we can verify that, if the norm $\|\tilde{\mathbf{m}} - \mathbf{m}\|_{L^\infty(\mathbb{R}^n)}$ is small enough, the element $\tilde{\mathbf{m}}$ is uniquely determined by w and $\mathbf{m}$ and, uniformly with respect to $\mathbf{m} \in \mathbb{P}(L)$, we have

$$\tilde{\mathbf{m}} - \mathbf{m} = \mathrm{w} + O(\|\,\mathrm{w}\,\|^2_{L^\infty(\mathbb{R}^n)}). \tag{9.21}$$

On the other hand, using the formula

$$\mathbb{P}_{\mathbf{m}}(t)\frac{d}{dt}\mathbb{P}_{\mathbf{m}(t)}s(t) = \mathbb{P}_{\mathbf{m}(t)}\frac{d}{dt}s(t) + \mathbb{D}(\mathbf{m}(t))[\mathbf{m}'(t)]s(t)$$

and equations for $\tilde{\mathbf{m}}$ and $\mathbf{m}$, we obtain that $\mathrm{w}(t)$ should satisfy

$$\begin{aligned}
&\mathbb{P}_{\mathbf{m}(t)}\tfrac{d}{dt}\,\mathrm{w}(t) - \mathbb{P}_{\mathbf{m}(t)}(\mathrm{f}'_{\mathbf{m}}(t,\mathbf{m}(t))(\tilde{\mathbf{m}}(t) - \mathbf{m}(t))) \\
&= \tilde{\mathrm{f}}(t,\tilde{\mathbf{m}}(t)) + \mathbb{P}_{\mathbf{m}(t)}[\mathrm{f}(t,\tilde{\mathbf{m}}(t)) - \mathrm{f}(t,\mathbf{m}(t)) - \mathrm{f}'(t,\mathbf{m}(t))(\tilde{\mathbf{m}}(t) - \mathbf{m}(t))] \\
&\quad + \mathbb{D}(\mathbf{m}(t))[\mathbf{m}'(t)](\tilde{\mathbf{m}}(t) - \mathbf{m}(t))
\end{aligned} \tag{9.22}$$

where $\mathbb{D}(\mathbf{m}) := \mathbb{P}_{\mathbf{m}} \circ \mathbb{P}'_{\mathbf{m}}$ is studied in Theorem 5.6. Using now (9.21) and the obvious fact that

$$\mathbb{D}(\mathbf{m})[\mathrm{w}_1]\,\mathrm{w}_2 \equiv 0, \quad \forall\, \mathrm{w}_i \in T_{\mathbf{m}}\mathbb{P}(L),$$

we see that (9.22) is indeed a small perturbation of the equation of variations and, consequently, the implicit function theorem is applicable. The rest of the proof repeats word by word the proof of Theorem 9.3 ad so omitted. Theorem 9.9 is proven. □

Thus, as in the previous case, hyperbolic sets are stable under the perturbations. The next corollary shows that the associated map $\mathcal{S} : \mathcal{H}^{tr} \to \tilde{\mathcal{H}}^{tr}$ is a homeomorphism in the local topology.

COROLLARY 9.10. *Let $\mathcal{H}^{tr}$ be a hyperbolic set of equation* (9.15)*, $\tilde{\mathrm{f}}$ be a sufficiently small perturbation and $\tilde{\mathcal{H}}^{tr}$ be the associated hyperbolic set of the perturbed equation. Assume in addition that the functions $\mathrm{f}(t,\mathbf{m})$, $\tilde{\mathrm{f}}(t,\mathbf{m})$ and $\mathrm{f}'(t,\mathbf{m})$ are continuous in the local topology. Then, the associated map $\mathcal{S} : \mathcal{H}^{tr} \to \tilde{\mathcal{H}}^{tr}$, $\mathcal{S}\mathbf{m} = \tilde{\mathbf{m}}$ is a homeomorphism in the topology of $L^\infty_{\mathrm{loc}}(\mathbb{R}^{n+1})$.*

The proof of this assertion is completely analogous to the proof of Corollary (9.5) and so is omitted.

To conclude the section, we study the relations between the hyperbolic sets of the full system (8.1) and of the reduced multi-pulse system (8.26) under the assumptions of Theorem 8.5. Indeed, let $\mathcal{H}^{tr}_{\mathrm{red}}$ be a hyperbolic set of trajectories of the reduced system (8.26) such that $\mathbf{m}(t) \in \mathbb{P}((1+\varepsilon)L)$, for every $t \in \mathbb{R}$ and $\mathbf{m} \in \mathcal{H}^{tr}_{\mathrm{red}}$. Then, due to the construction of the center manifold, the functions

$$u(t) := \mathbf{m}(t) + \mathbb{W}(t, \mathbf{m}(t)), \quad \mathbf{m} \in \mathcal{H}^{tr}_{\mathrm{red}} \tag{9.23}$$

solve the initial problem (8.1). Thus, the trajectory set $\mathcal{H}^{tr}_{\mathrm{full}}$ of the initial equation (8.1) associated with $\mathcal{H}^{tr}_{\mathrm{red}}$ is well-defined. The following corollary shows that $\mathcal{H}^{tr}_{\mathrm{full}}$ will be a hyperbolic trajectory set for equation (8.1).

COROLLARY 9.11. *Let the assumptions of Theorem 8.5 holds and let $\mathcal{H}^{tr}_{\mathrm{red}}$ be a hyperbolic set of the reduced equation* (8.26) *such that $\mathbf{m}(t) \in \mathbb{P}((1+\varepsilon)L)$ for all $t \in \mathbb{R}$ and $\mathbf{m} \in \mathcal{H}^{tr}_{\mathrm{red}}$ (where $\varepsilon > 0$ is the same as in Theorem 8.1). Then, the associated trajectory set $\mathcal{H}^{tr}_{\mathrm{full}}$ of the initial equation* (8.1) *will be also hyperbolic.*

PROOF. Let $\mathbf{m} \in \mathcal{H}^{tr}_{\mathrm{red}}$ be arbitrary and u be the associated trajectory of the full equation (8.1). We need to verify that u is hyperbolic in the sense of Definition 9.1.

According to Theorem 9.3 and Remark 9.4, it is sufficient to verify that, for every function $h \in L^\infty(\mathbb{R}^{n+1})$, the perturbed equation

$$\partial_t u + A_0 u + \Phi(u) - \mu R(t,u) = \nu h(t) \tag{9.24}$$

possesses a unique solution $\tilde{u} = \tilde{u}(\nu)$ if $\nu > 0$ is small enough and the following estimate holds:

$$\|\tilde{u}(t) - u(t)\|_{\mathbb{X}_b} \le C\nu \|h\|_{L^\infty(\mathbb{R}^{n+1})} \tag{9.25}$$

where the constant C is independent of ν, h and $t \in \mathbb{R}$.

We now recall that, due to Theorem (8.1), the perturbed equation (9.25) also possesses a center manifold reduction $\tilde{\mathbb{W}}(t,\mathbf{m})$ and the reduced equation reads

$$\frac{d}{dt}\tilde{\mathbf{m}}(t) = \tilde{\mathrm{f}}_h(t, \tilde{\mathbf{m}}(t), \tilde{\mathbb{W}}(t, \tilde{\mathbf{m}}(t))) \tag{9.26}$$

where the function $\tilde{\mathrm{f}}_h$ is obtained from f (which is, in turns, defined by (8.12)) by replacing μR by $\mu R + \nu h$. Moreover, due to Theorems 7.1 and 8.5, every solution of (9.24) which belongs to a sufficiently small neighborhood of $\mathbb{P}((1+\varepsilon)L)$ for all $t \in \mathbb{R}$ can be obtained from the reduced equation (9.26). On the other hand, due to Corollary 8.9, we have

$$\|\tilde{\mathbb{W}}(t,\cdot) - \mathbb{W}(t,\cdot)\|_{C^1(\mathbb{P}(L),\mathbb{X}_b)} \le C\nu\|h\|_{L^\infty(\mathbb{R}^{n+1})} \tag{9.27}$$

(where $\mathbb{W}(t,\cdot)$ is the center manifold reduction associated with $\nu = 0$) and, consequently,

$$\|\tilde{\mathrm{f}}_h(t,\cdot,\mathbb{W}(t,\cdot)) - \mathrm{f}(t,\cdot,\mathbb{W}(t,\cdot))\|_{C^1(\mathbb{P}(L),\mathbb{P}(L))} \le C\nu\|h\|_{L^\infty(\mathbb{R}^{n+1})}. \tag{9.28}$$

Thus, equation (9.26) is a small perturbation of equation (8.26) (which corresponds to the case $\nu = 0$). Since the initial trajectory $\mathbf{m} \in \mathcal{H}^{tr}_{\mathrm{red}}$ is assumed to be hyperbolic, then Theorem 9.9 gives that (9.26) is uniquely solvable in a small neighborhood of $\mathbf{m}$ (if $\nu > 0$ is small enough) and the associated solution $\tilde{\mathbf{m}}(t)$ satisfies

$$\|\tilde{\mathbf{m}}(t) - \mathbf{m}(t)\|_{L^\infty(\mathbb{R}^n)} \le C\nu\|h\|_{L^\infty(\mathbb{R}^{n+1})}. \tag{9.29}$$

Formula $\tilde{u}(t) = \tilde{\mathbf{m}}(t) + \tilde{\mathbb{W}}(t,\tilde{\mathbf{m}}(t))$ gives now the unique solution of (9.24) belonging to the small neighborhood of u and estimates (9.27) and (9.29) imply (9.25) which finishes the proof of hyperbolicity of u. The uniform hyperbolicity of $\mathcal{H}^{tr}_{\mathrm{full}}$ is an immediate corollary of the fact that the constant C in (9.25) depends only on the hyperbolicity constant $C_{\mathbf{m}}$ and is independent of the concrete choice of $\mathbf{m} \in \mathcal{H}^{tr}_{\mathrm{red}}$. Corollary 9.11 is proven. □

10. Multi-pulse evolution equations: asymptotic expansions

In this section, we compute the leading terms in the asymptotic expansion of the reduced ODEs on a center manifold as the distance $2L$ between pulses tends to ∞. Moreover, for some concrete examples of equations of the form (8.1), we give the explicit form of that terms. The following theorem can be considered as a main result of the section.

THEOREM 10.1. *Let the assumptions of Theorem 8.5 holds. Then, the reduced system (8.26) on a center manifold has the following structure:*

$$\frac{d}{dt}\mathbf{m} = \mathrm{f}_0(t,\mathbf{m}) + \bar{\mathrm{f}}(t,\mathbf{m}) \tag{10.1}$$

where

$$\mathrm{f}_0(t,\mathbf{m}) := \sum_{j\in\mathbb{N}} \mathbb{P}_{\Gamma_j}\left(\sum_{l\neq j} A_0 V_{\Gamma_l}\right) + \mu\sum_{j\in\mathbb{N}} \mathbb{P}_{\Gamma_j} R(t, V_{\Gamma_j}) \tag{10.2}$$

and the remainder $\bar{\mathrm{f}}$ satisfies

$$\|\bar{\mathrm{f}}(t,\cdot)\|_{C^1(\mathbb{P}(L),L^\infty(\mathbb{R}^n))} \le C_\varepsilon[\mathrm{e}^{-2(\alpha-\varepsilon)L} + \mu]^2 \tag{10.3}$$

where the constant C_ε depends on $\varepsilon > 0$, but is independent of L, μ and $\mathbf{m} \in \mathbb{P}((1+\varepsilon)L)$.

PROOF. We first recall that, due to the assumption $\mathbf{m} \in \mathbb{P}((1+\varepsilon)L)$ the cut-off operator $\mathrm{Cut}(\mathbf{m}) \equiv \mathrm{Id}$, the right-hand side $\tilde{f}$ in the pulse-equations (8.26) can be found by (8.12), i.e., the right-hand side of the reduced equations (8.26) reads

$$(10.4) \quad \tilde{f}(t,\mathbf{m},\mathbb{W}(t,\mathbf{m})) = \mathbb{M}(\mathbf{m},\mathbb{W}(t,\mathbf{m})(\mathbb{P}_{\mathbf{m}}(-\mathbb{F}(\mathbf{m}) - \Phi(\mathbb{W}(t,\mathbf{m}),\mathbf{m}) + \\ + \mu R(t,\mathbf{m}+\mathbb{W}(t,\mathbf{m}))) - \mathbb{S}(\mathbf{m})\mathbb{W}(t,\mathbf{m}))$$

Moreover, due to (8.25) and Lemma 8.1 the operator $\mathbb{M}(t,\mathbf{m}) := \mathbb{M}(\mathbf{m},\mathbb{W}(t,\mathbf{m}))$ satisfies

$$\|\mathbb{M}(t,\cdot) - \mathbb{P}_{\mathbf{m}}\|_{C^1(\mathbb{P}(L),\mathcal{L}(L^\infty(\mathbb{R}^n),L^\infty(\mathbb{R}^n)))} \le C_\varepsilon(\mathrm{e}^{-2(\alpha-\varepsilon)L} + \mu)$$

and, consequently (since f is also of order $C_\varepsilon(\mathrm{e}^{-2(\alpha-\varepsilon)L} + \mu)$, see Theorem 8.5), up to the remainder, we can set $\mathbb{M}(t,\mathbf{m}) = \mathrm{Id}$.

Furthermore, since the function $\Phi(v,\mathbf{m})$ is quadratic with respect to v, estimate (8.25) guarantees the $\Phi(\mathbb{W}(t,\mathbf{m}),\mathbf{m})$ also belongs to the remainder and, analogously, due to Theorem (5.8), the term $\mathbb{S}(\mathbf{m})\mathbb{W}(t,\mathbf{m})$ belongs to the remainder as well.

Finally, due to (8.25), $R(t,\mathbb{W}(t,\mathbf{m})+\mathbf{m}) = R(t,\mathbf{m}) + O(\mathrm{e}^{-2(\alpha-\varepsilon)L} + \mu)$ and, due to Corollary 5.4, $\|\mathbb{P}_{\mathbf{m}} - \sum_{j=1}^{\infty} \mathbb{P}_{\Gamma_j}\| \le C_\varepsilon \mathrm{e}^{-2(\alpha-\varepsilon)L}$. Thus, expression (10.4) can be represented as follows:

$$(10.5) \qquad \tilde{f}(t,\mathbf{m}) = -\sum_{j=1}^{\infty} \mathbb{P}_{\Gamma_j}\mathbb{F}(\mathbf{m}) + \mu \sum_{j=1}^{\infty} \mathbb{P}_{\Gamma_j} R(t,\mathbf{m}) + O([\mathrm{e}^{-2(\alpha-\varepsilon)L} + \mu]^2).$$

At the next step, using (2.28) and Lemma 3.12, we conclude that

$$|(R(t,\mathbb{V}_{\vec{\Gamma}}) - R(t,V_{\Gamma_j}),\psi^i_{\Gamma_j})| \le C_\varepsilon \sum_{l\ne j} \mathrm{e}^{-(\alpha-\varepsilon/2)|\xi_j-\xi_l|} \le C_\varepsilon \mathrm{e}^{-2(\alpha-\varepsilon)L}$$

and, consequently, up to the remainder the term $\sum_{j\in\mathbb{N}} \mathbb{P}_{\Gamma_j} R(t,\mathbf{m})$ can be replaced by $\sum_{j\in\mathbb{N}} \mathbb{P}_{\Gamma_j} R(t,V_{\Gamma_j})$. Thus, the theorem will be proved if we verify that

$$(10.6) \qquad \sum_{j\in\mathbb{N}} \mathbb{P}_{\Gamma_j}\mathbb{F}(\mathbf{m}) = -\sum_{j\in\mathbb{N}} \mathbb{P}_{\Gamma_j}(\sum_{l\ne j} A_0 V_{\Gamma_l}) + O(\mathrm{e}^{-4(\alpha-\varepsilon)L})$$

which is an immediate corollary of the following lemma.

LEMMA 10.2. *Let the above assumptions hold. Then, the following decomposition is valid:*

$$(10.7) \qquad (\mathbb{F}(\mathbb{V}_{\vec{\Gamma}}),\psi^i_{\Gamma_j}) = -\sum_{l\ne j}(A_0 V_{\Gamma_j},\psi^i_{\Gamma_j}) + [\mathcal{O}']^i_j(\vec{\Gamma})$$

where the remainder $\mathcal{O}'$ satisfies:

$$(10.8) \qquad \|\mathcal{O}'_j(\cdot)\|_{C^1(\mathbb{B}(L),\mathbb{R}^k)} \le C_\varepsilon \mathrm{e}^{-2(\alpha-\varepsilon)L} \mathrm{e}^{-(\alpha-\varepsilon)\,\mathrm{dist}'(\xi_j,\Xi(\vec{\Gamma}))}$$

where the constant C_ε depends on $\varepsilon > 0$, but is independent of L and $\vec{\Gamma}$.

PROOF. We give below only the estimate of the l^∞-norm of $\mathcal{O}'$ and the estimate of its $\vec{\Gamma}$-derivative can be obtained analogously. To this end, we first split $\mathbb{V}_{\vec{\Gamma}} := V_{\Gamma_j} + \mathbb{V}_{\vec{\Gamma}^j}$ and

$$(10.9) \qquad \mathbb{F}(\vec{\Gamma}) = [\Phi(V_{\Gamma_j} + \mathbb{V}_{\vec{\Gamma}^j}) - \Phi(V_{\Gamma_j}) - \Phi(\mathbb{V}_{\vec{\Gamma}^j})] + \mathbb{F}(\vec{\Gamma}^j)$$

where $\vec{\Gamma}^j = \{\Gamma_l\}_{l\neq j}$. Moreover, analogously to Lemma 4.8, we have

$$|\mathbb{F}(\vec{\Gamma}^j)|(x) \le C_\varepsilon \, e^{-(\alpha-\varepsilon)[\operatorname{dist}(x,\Xi^j)+\operatorname{dist}'(x,\Xi^j)]}$$

and, consequently, due to the triangle inequality, we have

$$(10.10)\quad |(\mathbb{F}(\vec{\Gamma}^j), \psi^i_{\Gamma_j})| \le C_\varepsilon \sup_{x\in\mathbb{R}^n} \{e^{-(\alpha-\varepsilon)|x-\xi_j|+\operatorname{dist}(x,\Xi^j)}\, e^{-(\alpha-\varepsilon)\operatorname{dist}'(x,\Xi)}\} \le \\ \le C_\varepsilon \, e^{\operatorname{dist}'(\xi_j,\Xi)}\, e^{-2(\alpha-\varepsilon)L}.$$

So, the term $(\mathbb{F}(\vec{\Gamma}^j), \psi^i_{\Gamma_j})$ belongs to the remainder. In order to estimate the first term in the right-hand side of (10.9), we use formula (4.34) with $v_1 = V_{\Gamma_j}$ and $v_2 := \mathbb{V}_{\vec{\Gamma}^j}$. Then, we have

$$(10.11)\quad \Phi(V_{\Gamma_j} + \mathbb{V}_{\vec{\Gamma}^j}) - \Phi(V_{\Gamma_j}) - \Phi(\mathbb{V}_{\vec{\Gamma}^j}) = \\ = \int_0^1\int_0^1 D^2_u\Phi(s_1V_{\Gamma_j} + s_2\mathbb{V}_{\vec{\Gamma}^j})\, ds_1\, ds_2[V_{\Gamma_j}, \mathbb{V}_{\vec{\Gamma}^j}] = \\ = \int_0^1\int_0^1 \{D^2_u\Phi(s_1V_{\Gamma_j} + s_2\mathbb{V}_{\vec{\Gamma}^j}) - D^2_u\Phi(s_1V_{\Gamma_j})\}\, ds_1\, ds_2[V_{\Gamma_j}, \mathbb{V}_{\vec{\Gamma}^j}] + \\ + \int_0^1\int_0^1 D^2_u\Phi(s_1V_{\Gamma_j})\, ds_1\, ds_2[V_{\Gamma_j}, \mathbb{V}_{\vec{\Gamma}^j}] := I_1 + I_2.$$

Using now assumptions (2.28) and Lemma 3.12, we verify

$$|(I_1, \psi^i_{\Gamma_j})| \le C_\varepsilon \sup_{x\in\mathbb{R}^n} \Big\{ \Big(\sum_{m\neq j} e^{-\alpha|x-\xi_m|} \Big)^2 e^{-2(\alpha-\varepsilon)|x-\xi_j|} \Big\} \le \\ \le C'_\varepsilon \sup_{x\in\mathbb{R}^n} \{e^{-2(\alpha-\varepsilon)\operatorname{dist}(x,\Xi^j)}\, e^{-2(\alpha-\varepsilon)|x-\xi_j|}\} \le C'_\varepsilon \, e^{-2(\alpha-\varepsilon)\operatorname{dist}'(\xi_j,\Xi)}$$

and, consequently, this term also belongs to the remainder. We now transform the term $(I_2, \psi^i_{\Gamma_j})$ as follows (using that $D_u\Phi(0) = 0$):

$$(10.12)\quad I_2 = \int_0^1\int_0^1 D^2_u\Phi(s_1V_{\Gamma_j})\, ds_1\, ds_2[V_{\Gamma_j}, \mathbb{V}_{\vec{\Gamma}^j}] = \\ = \int_0^1 D^2_u\Phi(s_1V_{\Gamma_j})\, ds_1[V_{\Gamma_j}, \mathbb{V}_{\vec{\Gamma}^j}] = [D_u\Phi(V_{\Gamma_j})]\mathbb{V}_{\vec{\Gamma}^j}$$

and, consequently, using the fact that the functions $\psi^i_{\Gamma_j}$ are eigenfunctions of the conjugate of the linearization of (2.1) at $u = V_{\Gamma_j}$, we finally deduce

$$(10.13)\quad (I_2, \psi^i_{\Gamma_j}) = ([D_u\Phi(\Gamma_j)]\mathbb{V}_{\vec{\Gamma}^j}, \psi^i_{\Gamma_j}) = -\sum_{l\neq j}(V_{\Gamma_l}, A_0^*\psi^i_{\Gamma_j}) = -\sum_{l\neq j}(A_0V_{\Gamma_l}, \psi^i_{\Gamma_j}).$$

Lemma 10.2 is proven. □

Consequently, Theorem 10.1 is also proven. □

The next corollary gives the analog of the proved theorem in local coordinates.

COROLLARY 10.3. *Let the assumptions of Theorem 10.1 hold. Then, the reduced pulse interaction equations* (8.26) *have the following form in the local coordinates* Γ^i_j *described in Definition 4.1 (see also Remark 4.5):*

$$\Pi(\Gamma_j)\frac{d}{dt}\Gamma_j = \sum_{m\neq j}\mathcal{F}([\Gamma_j]^{-1}\circ\Gamma_m) + \mu\mathcal{R}(t,\Gamma_j) + \mathcal{O}_j(t,\vec{\Gamma}),\quad j\in\mathbb{N} \tag{10.14}$$

where the matrix $\Pi(\Gamma)$ *is defined in* (2.37), *the functions* $\mathcal{F}$ *and* $\mathcal{R}$ *are defined via*

$$\mathcal{F}^i(\Gamma) := (V_\Gamma, A_0^*\psi^i)_{L^2(\mathbb{R}^n)},\quad \mathcal{R}^i(t,\Gamma) := (R(t,V_\Gamma),\psi^i_\Gamma)_{L^2(\mathbb{R}^n)},\quad i=1,\cdots,k, \tag{10.15}$$

the conjugate eigenfunctions ψ^i *are defined in* (2.22) *and* (2.23), *and the remainder* $\mathcal{O}$ *satisfies*

$$\|\mathcal{O}(t,\cdot)\|_{C^1(l^\infty,l^\infty)} \le C_\varepsilon[\mathrm{e}^{-2(\alpha-\varepsilon)L} + \mu]^2 \tag{10.16}$$

where the constant C *depends on* $\varepsilon>0$, *but is independent of* L *and* μ.

Indeed, taking the inner product of equation (10.1) with the function $\psi^i_j(\vec{\Gamma})$ and using the orthogonality relations (5.2), formula (2.37) and the obvious identity

$$(A_0V_{\Gamma_l},\psi^i_{\Gamma_j}) = (A_0\mathcal{T}_{\Gamma_l}V,\mathcal{T}_{\Gamma_j}\psi^i) = \\ = (\mathcal{T}_{[\Gamma_j^{-1}}A_0\mathcal{T}_{\Gamma_l}V,\psi^i) = (A_0\mathcal{T}_{[\Gamma_j]^{-1}\cdot\Gamma_l}V,\psi^i) = (A_0V_{[\Gamma_j]^{-1}\cdot\Gamma_j},\psi^i),$$

we obtain equations (10.14) in any local coordinates belonging to the atlas described in Remark 4.5.

REMARK 10.4. We see that, due to Theorem 10.1, up to the terms of order $[\mathrm{e}^{-2(\alpha-\varepsilon)}\,L+\mu]^2$, the reduced system of ODEs has the form

$$\Pi(\Gamma_j)\frac{d}{dt}\Gamma_j = \sum_{m\neq j}\mathcal{F}([\Gamma_j]^{-1}\circ\Gamma_m) + \mu\mathcal{R}(t,\Gamma_j),\quad j\in\mathbb{N}. \tag{10.17}$$

Here, the terms $\mathcal{F}([\Gamma_j]^{-1}\circ\Gamma_m)$, $m\neq j$ can be interpreted as a pairwise "tail"-interaction between pulses V_{Γ_j} and V_{Γ_m} and the term $\mu\mathcal{R}(t,\Gamma_j)$ is a "self-interaction" of the jth pulse appeared due to the perturbation of the initial equation (2.1). We also note that, formally, equation (10.17) contains infinitely many interaction terms $\mathcal{F}([\Gamma_j]^{-1}\circ\Gamma_m)$. Nevertheless, only finite number of them are factually larger than $[\mathrm{e}^{-2(\alpha-\varepsilon)}\,L+\mu]^2$ and, consequently, the others can be included into the remainder $\mathcal{O}$. To be more precise, arguing as in Lemma 10.2, it is easy to verify that

$$\|\mathcal{F}([\Gamma_j]^{-1}\circ\Gamma_m)\| \le C_\varepsilon\,\mathrm{e}^{-(\alpha-\varepsilon)|\xi_j-\xi_m|} \tag{10.18}$$

and therefore, due to Lemma 3.12, all the terms $\mathcal{F}([\Gamma_j]^{-1}\circ\Gamma_m)$ with $|\xi_j-\xi_m|\ge 4L$ can be also interpreted as a part of remainder. Then, system (10.17) reads

$$\Pi(\Gamma_j)\frac{d}{dt}\Gamma_j = \sum_{m\in\mathbb{N}_{\vec{\Gamma}}(j)}\mathcal{F}([\Gamma_j]^{-1}\circ\Gamma_m) + \mu\mathcal{R}(t,\Gamma_j),\quad j\in\mathbb{N} \tag{10.19}$$

where $\mathbb{N}_{\vec{\Gamma}}(j) := \{m\in\mathbb{N},\ \ |\xi_j-\xi_m|<4L\}$. It is also worth to note that the number of points in $\mathbb{N}_{\vec{\Gamma}}(j)$ is finite and uniformly bounded

$$\#\mathbb{N}_{\vec{\Gamma}}(j) \le C_0 = C_0(n) \tag{10.20}$$

where the constant C_0 depends only on n (and is independent of $\vec{\Gamma}$, L and j). In particular, in one-dimensional case $n=1$, $C_0=2$ and consequently, for every given pulse V_{Γ_j}, we have at most two neighboring pulses on the distance less than $4L$ from it. Moreover, in one-dimensional case we have a natural ordering on the set of pulses: $\Gamma_i < \Gamma_j$ if and only if $\xi_i < \xi_j$ which *preserves* under the evolution inside of the center manifold. Thus, after the renumeration of the pulses in the increasing order, equations (10.19) read:

$$\Pi(\Gamma_j)\frac{d}{dt}\Gamma_j = \mathcal{F}([\Gamma_j]^{-1}\circ\Gamma_{j-1}) + \mathcal{F}([\Gamma_j]^{-1}\circ\Gamma_{j+1}) + \mu\mathcal{R}(t,\Gamma_j), \quad j\in\mathbb{Z} \tag{10.21}$$

which are usual for the one-dimensional theory, see [**Ei2002, San93, San02**].

In contrast to that, in the multidimensional case $n \geq 2$, we do not have any canonical ordering of the pulses and, moreover, the neighboring pulses to V_{Γ_j} (i.e. the pulses V_{Γ_m}, $m \in \mathbb{N}_{\vec{\Gamma}}(j)$) also can change under the time evolution (the set $\mathbb{N}_{\vec{\Gamma}(t)}(j)$ can depend on t). That is the reason why, we retain in (10.14) *all of the* "tail"-interactive terms.

We now going to consider several examples of equations of the form (8.1) and give some explicit formulae for computing the pairwise interaction function $\mathcal{F}(\Gamma)$. We start with the case of a spherically symmetric pulse and gradient equations.

EXAMPLE 10.5. Let us assume that initial pulse $V(x)$ of equation (2.1) is spherically symmetric and that the linearized operator $\mathcal{L}_V := A_0 + D_u\Phi(V(x))$ is self-adjoint:

$$V(x) := V(r^2), \quad r := |x|, \quad [\mathcal{L}_V]^* = \mathcal{L}_V. \tag{10.22}$$

Then, we obviously have n-eigenfunctions of operator $\mathcal{L}_V$ generated by derivatives of the initial pulse V

$$\phi^i(x) := -\partial_{x_i}V(x) = -2x_iV'(r^2), \quad i = 1,\cdots,n. \tag{10.23}$$

Furthermore, we assume that these functions generate the whole kernel of $\mathcal{L}_V$, then this pulse satisfies all of the assumptions of Section 1 with the symmetry group of spatial translations $G := \mathbb{R}^n$, $\Gamma := \xi \in \mathbb{R}^n$ and

$$(\mathcal{T}_\xi u)(x) := u(x-\xi). \tag{10.24}$$

In that case, we have $V_\Gamma(x) = V_\xi(x) := V(x-\xi)$ and $\partial_{\xi_i}V_\xi = -\partial_x V(\cdot - \xi) = \phi^i(\cdot-\xi) = \phi^i_\xi$. According to (2.37), this gives

$$\Pi(\xi) = \mathrm{Id}\,. \tag{10.25}$$

Moreover, it is not difficult to verify that $(\phi^i,\phi^j) = 0$ for $i \neq j$ and, since, the operator $\mathcal{L}_V$ is assumed to be self-adjoint, the conjugate functions ψ^i have the following form:

$$\psi^i(x) := c^{-1}\phi^i(x), \quad c^2 := \|\phi^i\|^2_{L^2(\mathbb{R}^n)} = \frac{1}{n}\int_{\mathbb{R}^n} r^2[V'(r^2)]^2\,dx \tag{10.26}$$

(the multiplier c^{-1} is chosen in order to satisfy (2.23)). Inserting these formulae to the definition of $\mathcal{F}(\xi)$, we establish that, in this case the interaction function has a gradient structure $\mathcal{F}(\xi) = \nabla_\xi\mathcal{G}(\xi)$ with the potential

$$\mathcal{G}(\xi) := c^{-1}\int_{\mathbb{R}^n} V(x-\xi).A_0V(x)\,dx. \tag{10.27}$$

where $u.v$ means the standard inner product in $\mathbb{R}^m$. As a particular example of equations of that type one can consider the scalar reaction-diffusion equation

$$\partial_t u = a\Delta_x u - f(u). \tag{10.28}$$

The existence theorems for a spherically symmetric pulse $V(r^2)$ in the gradient case (the so-called ground state solution for this equation (under some natural assumptions on the nonlinearity f) can be proven e.g. by variational methods, see in [**BeL83**]. Moreover, the computation of the interaction function $\mathcal{F}$ for more complicated non-gradient equations with spherically symmetric pulse in two dimensions related with nonlinear optics (including the asymptotics for $\mathcal{G}(\xi)$ as $\xi \to \infty$) and the examples of spectrally stable pulses can be found in [**TVM03**], [**VMSF02**].

The next example will be the so-called generalized 1D Swift-Hohenberg equation which is factually a particular case of Example 10.5.

EXAMPLE 10.6. Let us consider the following problem in $x \in \mathbb{R}$:

$$\partial_t u + (\partial_x^2 + 1)^2 u + \beta^2 u + f(u) = 0, \quad f(0) = f'(0) = 0. \tag{10.29}$$

Here $A_0 := (\partial_x^2 + 1)^2 + \beta^2$ and, obviously, (10.29) has a gradient structure, so we are under the assumptions of Example 10.5. Our task now is to find the explicit form of function $\mathcal{G}(\xi)$ and/or its asymptotics as $\xi \to \infty$. Our assumption is that equation (10.29) possesses a nondegenerate *symmetric* pulse equilibrium $V(x)$, i.e.

$$V(x) = V(-x). \tag{10.30}$$

It is known that such pulses exist, e.g. for the cubic nonlinearity

$$f(u) := u^3 + \kappa u^2, \quad \kappa \in \mathbb{R} \tag{10.31}$$

in some region of parameters β and κ, see [**BGL97, GlL94**].

Let $\lambda \in \mathbb{C}$, $\alpha := \operatorname{Re}\lambda > 0$ solve the characteristic equation:

$$(\lambda^2 + 1)^2 + \beta^2 = 0. \tag{10.32}$$

Then, the pulse $V(x)$ has the following asymptotics as $x \to \pm\infty$:

$$V(x) = \operatorname{Re}\{V_0 \, e^{\mp\lambda x}\} + V_r(x), \quad |V_r(x)| \le C\, e^{-2\alpha|x|} \tag{10.33}$$

for some $V_0 \in \mathbb{C}$, $|V_0| \neq 0$ (here we have implicitly used the assumption that the pulse $V(x)$ is symmetric. We are now going to simplify expression (10.27) for $\mathcal{G}$ using asymptotics (10.33). To this end, we also need the following version of Green's formula for the operator A_0 which can be easily verified by integration by parts:

$$\int_0^\infty [g_1(x).A_0 g_2(x) - A_0 g_1(x).g_2(x)]\, dx = K_0(g_1, g_2) := -[g_1(0)g_2'''(0) - \\ - g_1'(0)g_2''(0) + g_1''(0)g_2'(0) - g_1'''(0)g_2(0) + 2g_1(0)g_2'(0) - 2g_1'(0)g_2(0)]. \tag{10.34}$$

Using (10.33) and (10.34) and the fact that $A_0\,e^{\lambda x}\equiv 0$, we have

$$(10.35)\quad (V(\cdot-\xi),A_0V(\cdot)) = \\ =\int_{-\infty}^{0} V(x-\xi).A_0V(x)\,dx+\int_{0}^{\infty} V(x-\xi).A_0V(x)\,dx= \\ =\int_{-\infty}^{0} V(x-\xi).A_0[V(x)-\mathrm{Re}\;V_0\,e^{\lambda x}]\,dx+ \\ +\int_{0}^{\infty} V(x-\xi).A_0[V(x)-\mathrm{Re}\;V_0\,e^{-\lambda x}]\,dx=\int_{-\infty}^{0} A_0V(x-\xi).V_r(x)\,dx+ \\ +\int_{0}^{\infty} A_0V(x-\xi).V_r(x)\,dx-K_0(V(x-\xi),V(x)-\mathrm{Re}\;V_0\,e^{\lambda x})+ \\ +K_0(V(x-\xi),V(\cdot)-\mathrm{Re}\;V_0\,e^{-\lambda x})=\int_{-\infty}^{0} A_0V(x-\xi).V_r(x)\,dx+ \\ +\int_{0}^{\infty} A_0V(x-\xi).V_r(x)\,dx+\mathrm{Re}\{V_0K_0(V(x-\xi),e^{-\lambda x}-e^{\lambda x})\}= \\ =\int_{R} A_0V(x-\xi).V_r(x)\,dx+2\,\mathrm{Re}\{V_0\lambda(V(-\xi)\lambda^2+V''(-\xi)+2V(-\xi))\}.$$

Moreover, due to (10.34), we have $|V_r(x)|\le C\,e^{-2\alpha|x|}$ and

$$(10.36)\qquad |A_0V(x-\xi)|=|f(V(x-\xi)|\le C\,e^{-2\alpha|x-\xi|}.$$

Therefore,

$$\Big|\int_{R} A_0V(x-\xi).V_r(x)\,dx\Big|\le C_\varepsilon\,e^{-2(\alpha-\varepsilon)|\xi|}\le C_\varepsilon\,e^{-4(\alpha-\varepsilon)L}.$$

Thus, up to the terms of order $e^{-4(\alpha-\varepsilon)L}$, we have

$$(10.37)\qquad \mathcal{G}(\xi)=2c^{-1}\,\mathrm{Re}\{V_0\lambda(V(-\xi)\lambda^2+V''(-\xi)+2V(-\xi))\}.$$

Furthermore, since the function $V(x)$ is even, then the potential $\mathcal{G}(\xi)$ is also even with respect to ξ. So, without loss of generality, we may assume that $\xi>0$. Inserting now $V(-\xi)=1/2[V_0\,e^{-\lambda\xi}+\bar V_0\,e^{-\bar\lambda\xi}]$ (up to the terms of order $e^{-2\alpha|\xi|}$), we deduce that

$$(10.38)\quad V(-\xi)\lambda^2+V''(-\xi)+2V(-\xi):=V_0\,e^{-\lambda\xi}[\lambda^2+1]+ \\ +1/2\bar V_0\,e^{-\bar\lambda\xi}[\lambda^2+\bar\lambda^2+2]=V_0 i\beta\,e^{-\lambda\xi}$$

where we have used that $\lambda^2+1=i\beta$. Thus, up to the terms of order $e^{-4(\alpha-\varepsilon)L}$, we have

$$(10.39)\qquad \mathcal{G}(\xi)=2c^{-1}\beta\,\mathrm{Re}\{iV_0^2\lambda\,e^{-\alpha\xi-i\,\mathrm{Im}\,\lambda\xi}\}=M_0\,e^{-\alpha|\xi|}\sin(\mathrm{Im}\,\lambda|\xi|+\phi_0)$$

where $M_0=2c^{-1}|V_0|^2\beta|\lambda|=2c^{-1}|V_0|^2\beta\sqrt{1+\beta^2}>0$ if $\beta\ne 0$ and $\phi_0\in\mathbb{R}$ is some real number. Finally, returning to the function $\mathcal{F}$, we have, up to the terms of order $e^{-4(\alpha-\varepsilon)L}$

$$(10.40)\qquad \mathcal{F}(\xi)=M_0'\,\mathrm{sgn}(\xi)\,e^{-\alpha|\xi|}\sin(\omega|\xi|+\phi_0')$$

where $M_0'>0$, $\omega:=\mathrm{Im}\,\lambda\ne 0$ and $\phi_0'\in\mathbb{R}$. The reduced system of ODEs on the center manifold associated with equation (10.29) reads (up to the terms of order

$e^{-2(\alpha-\varepsilon)L})$

$$\frac{d}{dt}\xi_j = M_0'[e^{-\alpha|\xi_{j-1}-\xi_j|}\sin(\omega|\xi_{j-1}-\xi_j|+\phi_0') - e^{-\alpha|\xi_{j+1}-\xi_j|}\sin(\omega|\xi_{j+1}-\xi_j|+\phi_0')],$$

where $j \in \mathbb{Z}$ and the pulses V_{ξ_j} are numerated in the increasing order.

We consider also the perturbed version of equation (10.29) (which will be used in the next section for constructing the example of space-time chaos):

$$\partial_t u + (\partial_x^2+1)^2 u + \beta^2 u + f(u) = \mu h(t,x), \quad f(0) = f'(0) = 0 \tag{10.41}$$

for some $h \in L^\infty(\mathbb{R}^{n+1})$. Then, according to Theorem 10.1, the reduced ODEs on the center manifold read (up to the terms of order $[e^{-2(\alpha-\varepsilon)L} + \mu]^2$):

$$\frac{d}{dt}\xi_j = M_0'[e^{-\alpha|\xi_{j-1}-\xi_j|}\sin(\omega|\xi_{j-1}-\xi_j|+\phi_0') - \\ - e^{-\alpha|\xi_{j+1}-\xi_j|}\sin(\omega|\xi_{j+1}-\xi_j|+\phi_0')] + \mu\mathcal{R}(t,\xi_j) \tag{10.42}$$

with $\mathcal{R}(\xi) := -c^{-1}\int_R h(t,x)\partial_x V(x-\xi)\,dx$.

REMARK 10.7. We see from (10.42) that if $\operatorname{dist}(\xi_j,\xi_{j+1}) \sim 2L$ then the term $\mathcal{F}(\xi_{j+1}-\xi_j)$ is of order $e^{-2\alpha L}$ and, therefore, gives indeed the leading term of the asymptotic expansion as $L \to \infty$. It is also worth to emphasize that the form of the interaction functions in (10.42) depends only on the properties of the linear part of equation (10.29) and is independent of the nonlinearity f. So, the concrete choice of f determines only the amplitude M_0' and the phase ϕ_0' in equations (10.42). It was however crucial to verify that $M_0' \neq 0$ (otherwise, the first term of the expansion would vanish identically and one need to compute the further term of the asymptotics). The fact that the frequency $\omega \neq 0$ is also very important for what follows, see the next section.

We conclude this section by considering the generalized 1D Ginzburg-Landau equation.

EXAMPLE 10.8. Let us consider the following problem:

$$\partial_t u - (1+i\kappa)\partial_x^2 u + \gamma u + u f(|u|^2), \quad f(0) = 0 \tag{10.43}$$

where $u(t,x) = u_1(t,x) + iu_2(t,x)$ is an unknown complex-valued function, $f : \mathbb{R} \to \mathbb{C}$ is a given nonlinearity and $\kappa \in \mathbb{R}$ and $\gamma \in \mathbb{C}$ are parameters. This equation possesses a natural group of symmetries $G := \mathbb{R}^1 \times S^1$, $\Gamma := (\xi,\phi)$ and

$$(\mathcal{T}_{(\xi,\phi)}u)(x) := e^{i\phi}\,u(x-\xi), \quad \phi,\xi \in \mathbb{R} \tag{10.44}$$

which are isometries with respect to the inner product $(u,v) := \operatorname{Re}\int_R u(x)\bar{v}(x)\,dx$.

We assume that equation (10.43) possesses a homoclinic equilibrium solution $V(x)$ to $u \equiv 0$ which is symmetric with respect to x, i.e. $V(-x) = V(x)$. Then, analogously to (10.33), this solution necessarily has the following asymptotics as $x \to \pm\infty$:

$$V(x) = V_0\,e^{\mp\lambda x} + V_r(x), \quad x \to \pm\infty, \quad \text{where } \lambda := \sqrt{\frac{\gamma}{1+i\kappa}} \tag{10.45}$$

with $\lambda = \alpha + i\omega$, $\alpha > 0$ and, as in the previous example $|V_r(x)| \le C\,e^{-2\alpha|x|}$. See [**AfM99, AfM01**] for existence of single and multi-pulse equilibria for (10.43) as well as for the Poiseuille profile. See also, [**ASCT01**] for numerical and experimental observations of different pulses in (10.43).

In contrast to the previous example, we now have two-dimensional group of symmetries and, consequently, the dimension of the kernel for the nondegenerate pulse $V(x)$ is equal two and the corresponding eigenfunctions are

$$\phi^1_\Gamma(x) := -\operatorname{e}^{i\phi} V'(x-\xi), \quad \phi^2_\Gamma(x) := i\operatorname{e}^{i\phi} V(x-\xi). \tag{10.46}$$

Since G_0 is Abelian, we again have $\Pi(\Gamma) = \mathrm{Id}$. Moreover, the function ϕ^1 is odd with respect to x and ϕ^2 is even. Consequently $(\phi^1, \phi^2) = 0$. We however note, that our equation is now not gradient, so we need to introduce also the adjoint eigenfunctions $\psi^i_\Gamma(x)$ (which do not coincide as before with $\phi^i_\Gamma(x)$ and should be computed by solving the adjoint equation $\mathcal{L}^*_{V_\Gamma}\psi^i_\Gamma = 0$). Without loss of generality, we may assume that $\psi^1(x)$ is odd and $\psi^2(x)$ is even, so the orthogonality relations $(\phi^1, \psi^2) = (\phi^2, \psi^1) = 0$ are automatically satisfied and the rest of the (2.23) ($(\phi^1, \psi^1) = (\phi^2, \psi^2) = 1$) can be obtained by scaling. Furthermore, analogously to (10.45), we have

$$\psi^l(x) = (\pm 1)^l \bar{\Psi}^l_0 \operatorname{e}^{\pm\bar{\lambda}x} + \psi^l_r(x), \quad x \to \pm\infty \tag{10.47}$$

for some $\Psi^l \in \mathbb{C}$ and $l = 1, 2$. According to Theorem 10.1, we need to compute the following quantities:

$$\mathcal{F}^l(\xi, \phi) = \operatorname{Re}\{\operatorname{e}^{i\phi}\int_R V(x-\xi)A_0\bar{\psi}^l(x)\,dx\}, \quad l = 1, 2 \tag{10.48}$$

with $A_0 := -(1+i\kappa)\partial_x^2 + \gamma$. Inserting now the asymptotic expansions (10.47) into the right-hand side of (10.48) and arguing exactly as in (10.35), we infer

$$\begin{aligned}\int_R V(x-\xi)A_0\bar{\psi}^l(x)\,dx = (1-i\kappa)\Psi^l_0\{&V(x-\xi)\partial_x((-1)^l\operatorname{e}^{\lambda x} - \operatorname{e}^{-\lambda x}) - \\ &- V'(x-\xi)((-1)^l\operatorname{e}^{\lambda x} - \operatorname{e}^{-\lambda x})\}\big|_{x=0} + O(\operatorname{e}^{-2\alpha|\xi|})\end{aligned} \tag{10.49}$$

and, consequently,

$$\begin{cases} \mathcal{F}^1(\xi, \phi) = 2\operatorname{Re}\{(1-i\kappa)\operatorname{e}^{i\phi}\Psi^1_0 V'(-\xi)\} + O(\operatorname{e}^{-2\alpha|\xi|}), \\ \mathcal{F}^2(\xi, \phi) = 2\operatorname{Re}\{(1-i\kappa)\operatorname{e}^{i\phi}\Psi^2_0 \lambda V(-\xi)\} + O(\operatorname{e}^{-2\alpha|\xi|}). \end{cases} \tag{10.50}$$

Moreover, since $V(x)$ and $\psi^2(x)$ are even and ψ^1 is odd, then $\mathcal{F}^1$ and $\mathcal{F}^2$ are odd and even with respect to ξ respectively. So, without loss of generality, we may assume that $\xi > 0$. Then, due to (10.45), we have

$$\begin{cases} \mathcal{F}^1(\xi, \phi) = 2\operatorname{Re}\{(1-i\kappa)\operatorname{e}^{i\phi}\Psi^1_0 V_0 \lambda \operatorname{e}^{-\lambda\xi}\} + O(\operatorname{e}^{-2\alpha|\xi|}), \\ \mathcal{F}^2(\xi, \phi) = 2\operatorname{Re}\{(1-i\kappa)\operatorname{e}^{i\phi}\Psi^2_0 V_0 \lambda \operatorname{e}^{-\lambda\xi}\} + O(\operatorname{e}^{-2\alpha|\xi|}). \end{cases} \tag{10.51}$$

Thus, up to the terms of order $\operatorname{e}^{-4\alpha L}$ (which can be considered as a part of the remainder), we have

$$\begin{cases} \mathcal{F}^1(\xi, \phi) = M^1_0 \operatorname{sgn}\{\xi\}\operatorname{e}^{-\alpha|\xi|}\sin(\phi + \omega|\xi| + \theta_1), \\ \mathcal{F}^2(\xi, \phi) = M^2_0 \operatorname{e}^{-\alpha|\xi|}\sin(\phi + \omega|\xi| + \theta_2), \end{cases} \tag{10.52}$$

where $M_0^l := |2(1-i\kappa)\Psi_0^l V_0 \lambda| \neq 0$ and $\theta_l \in \mathbb{R}$. Finally, according to (10.21) the ODE on the center manifold has the following form:

$$\begin{aligned} \frac{d}{dt}\xi_j &= M_0^1[\mathrm{e}^{-\alpha(\xi_{j+1}-\xi_j)}\sin(\omega(\xi_{j+1}-\xi_j)-\phi_{j+1}+\phi_j+\theta_1)- \\ &\qquad -\mathrm{e}^{-\alpha(\xi_j-\xi_{j-1})}\sin(\omega(\xi_j-\xi_{j-1})+\phi_j-\phi_{j-1}+\theta_1)], \\ \frac{d}{dt}\phi_j &= M_0^2[\mathrm{e}^{-\alpha(\xi_{j+1}-\xi_j)}\sin(\omega(\xi_{j+1}-\xi_j)-\phi_{j+1}+\phi_j+\theta_2)+ \\ &\qquad +\mathrm{e}^{-\alpha(\xi_j-\xi_{j-1})}\sin(\omega(\xi_j-\xi_{j-1})+\phi_j-\phi_{j-1}+\theta_2)], \end{aligned} \tag{10.53}$$

where $j \in \mathbb{Z}$ and the pulses are numerated in an increasing order.

REMARK 10.9. The existence of the pulse solution in Example 10.6 is verified analytically or/and numerically for wide class of equation (10.43) including the classical Ginzburg-Landau equation with cubic nonlinearity, quintic Ginzburg-Landau equation (where the above pulse can be spectrally stable) and even for the case of non-polynomial nonlinearity f arising in laser dynamics, see [**AfM99, VFK*99, VKR01, VMSF02**]. Formal derivation of equations (10.53) can be found also in [**SkV02**].

11. An application: spatio-temporal chaos in periodically perturbed Swift-Hohenberg equation

In this concluding section we show how the multi-pulse center manifold reduction can be applied in order to prove that the concrete equations of mathematical physics possess the so-called space-time chaos. One of the most natural mathematical model for that phenomena was suggested by Sinai and Bunimovich [**BuS88**] (see also [**PeS88**] and [**PeS91**]). A discrete lattice dynamical system consisting of a $\mathbb{Z}^n$-grid of weakly coupled chaotic in time oscillators has been considered there. Then, without taking into account the coupling, the above system is just a $\mathbb{Z}^n$-Cartesian product of the basic chaotic oscillator. In particular, if this chaotic oscillator contains a hyperbolic set $\mathbb{G}_0$ then the uncoupled system will contain an infinite-dimensional hyperbolic set $\mathbb{G} \sim (\mathbb{G}_0)^{\mathbb{Z}^n}$. Therefore, due to the structure stability theorem for hyperbolic sets, the coupled system also possesses a hyperbolic set homeomorphic to $\mathbb{G}$ if the constant of interaction is small enough.

We also recall that the simplest model example of a hyperbolic set $\mathbb{G}_0$ is the one generated by a single transversal homoclinic orbit. Then it is naturally homeomorphic to the Bernoulli scheme $\mathcal{M}^1 := \{0,1\}^{\mathbb{Z}}$. In this case, the infinite-dimensional hyperbolic set $\mathbb{G}$ of the whole coupled system will be homeomorphic to the multi-dimensional Bernoulli scheme $\mathcal{M}^{n+1} := \{0,1\}^{\mathbb{Z}^{n+1}} = (\mathcal{M}^1)^{\mathbb{Z}^n}$ (see Definition 11.1 below). Thus, the multidimensional Bernoulli scheme $\mathcal{M}^{n+1}$ can be considered as a natural model for describing the space-time chaos phenomena in dissipative systems in unbounded domains.

It is worth to note that the existence of the above hyperbolic sets in concrete equations of mathematical physics seems to be a highly nontrivial problem and for a long time there were not any example of a dissipative PDE where this existence were rigorously verified. The first examples of such PDEs has been recently suggested in [**MiZ04**] where the above hyperbolic sets were found on the attractors of some special space-time periodic reaction-diffusion equations in $\mathbb{R}^n$. However, the nonlinear interaction function for that examples is very complicated and constructed artificially, i.e., it is rather far from the concrete nonlinearities arising in mathematical physics.

In this section, we present an alternative method of constructing such hyperbolic sets based on the analysis of weak interaction of an infinite grid of pulses and the center manifold reduction obtained in previous sections. We believe that this method can be applied for many concrete equations of mathematical physics. We illustrate this method on model example of the generalized 1D Swift-Hohenberg equation considered in Example 10.6. Further applications will be considered in forthcoming papers.

Thus, we consider the space-time periodically forced Swift-Hohenberg equation:

$$\partial_t u + (\partial_x^2 + 1)^2 u + \beta^2 u + f(u) = \mu h(t,x), \quad f(0) = f'(0) = 0 \tag{11.1}$$

where $h(t,x) = h_\mu(t,x)$ is some special space-time periodic external force and μ is a small parameter. Our main assumption concerning the non-perturbed equation (with $\mu = 0$) is that equation possesses a nondegenerate *pulse* $V(x)$ which satisfy the assumptions of the center manifold theorem proved above. Then, we show that under some special forcing h *pairs* of weakly interacting pulses generate chaotic oscillators in time with the hyperbolic set $\mathbb{G}_0 = \mathcal{M}^1$. Considering after that a $\mathbb{Z}$-grid of such pairs of pulses and using the center manifold reduction, we verify the existence of a hyperbolic set $(\mathbb{G}_0)^{\mathbb{Z}} = \mathcal{M}^2$.

In order to formulate the precise results, we first recall the notion of a Bernoulli scheme.

DEFINITION 11.1. Let $\mathcal{M}^n$ be a Cartesian product $\{0,1\}^{\mathbb{Z}^n}$ endowed by the Tikhonov topology (by definition, this means that $\mathcal{M}^n$ consists of functions $v : \mathbb{Z}^n \to \{0,1\}$ endowed by the local topology). Obviously, the group of shifts $\{\mathcal{S}_l, l \in \mathbb{Z}^n\}$ naturally acts on $\mathcal{M}^n$ via

$$(\mathcal{S}_l v)(m) := v(l+m), \quad l, m \in \mathbb{Z}^n, \quad v \in \mathcal{M}^n. \tag{11.2}$$

The set $\mathcal{M}^n$ and the group $\mathcal{S}_l$ acting on it are usually called Bernoulli scheme and Bernoulli shifts respectively. We also introduce the natural one-parametric subgroups $\{S_l^i, l \in \mathbb{Z}\}$, $i = 1, \cdots, n$ by the following expression:

$$\mathcal{S}_l^i := \mathcal{S}_{l\vec{e}_i}, \quad l \in \mathbb{Z} \tag{11.3}$$

where e_i is the ith coordinate vector in $\mathbb{R}^n$.

We are now ready to formulate the main result of this section which establishes the existence of an infinite-dimensional hyperbolic set of the above form for the generalized Swift-Hohenberg equation.

THEOREM 11.2. *Let the nonlinearity f and the parameter β in equation* (11.1) *be such that the non-perturbed equation (with $\mu = 0$) possesses a homoclinic pulse satisfying the assumptions of Theorem 8.5. Then, for every sufficiently small $\mu > 0$ there exist two positive numbers $P_t = P_t(\mu)$ and $P_x = P_x(\mu)$ (tending to infinity as $\mu \to 0$) and a smooth external force $h = h_\mu$ which is P_t-periodic in time and P_x-periodic in space (and is uniformly bounded as $\mu \to 0$, i.e. $\|h_\mu\|_{C_b^1(\mathbb{R}^n)} \le C$) such that the forced Swift-Hohenberg equation* (11.1) *possesses a hyperbolic set $\mathcal{H}^{tr} \subset W_b^{(1,4),p}(\mathbb{R}^2)$ of trajectories (in the sense of Definition 9.1). Moreover, there exists a one-to-one map*

$$\kappa : \mathcal{M}^2 \to \mathcal{H}$$

which satisfies the following commutation relations:

$$T^1_{lP_t} \circ \kappa = \kappa \circ \mathcal{S}^1_l, \quad T^2_{lP_x} \circ \kappa = \kappa \circ \mathcal{S}^2_l, \quad l \in \mathbb{Z} \tag{11.4}$$

where T_l is a space-time translation group, i.e. $(T^1_s u)(t,x) := u(t+s,x)$ and $(T^2_s u)(t,x) := u(t,x+s)$. Furthermore, the map κ is a homeomorphism if the set $\mathcal{H}^{tr}$ is endowed by the local topology of $W^{(1,4),p}_{\mathrm{loc}}(\mathbb{R}^2)$.

PROOF. As already mentioned, in order to prove the theorem, we consider a special grid of pair of pulses. To this end, for every sufficiently large L, we define the numbers $L_j \in \mathbb{R}$, $j \in \mathbb{Z}$, by the following recursive procedure:

$$L_0 = 0, \quad L_{2j+1} - L_{2j} = 2L, \quad L_{2j} - L_{2j-1} = 4L, \quad j \in \mathbb{Z}. \tag{11.5}$$

The numbers L_j will play the role of nodes of our grid of pulses. To be more precise, we seek for the pulse configuration $\vec{\xi}(t) = \{\xi_j(t)\}_{j\in\mathbb{Z}}$ satisfying

$$|L_j - \xi_j(t)| \le C, \quad t \in \mathbb{R}. \tag{11.6}$$

Keeping in mind (11.6), we introduce the new pulse coordinates $\bar{\xi}_j := \xi_j - L_j$ which should be of order 1 as $L \to \infty$. Moreover, without loss of generality, we may assume that $2\omega L + \theta = 2\pi k$ for some $k \in \mathbb{N}$. Then, according to Theorem 10.1 and computations of Example 10.6, the reduced ODE on the center manifold reads

$$\begin{cases} \dfrac{d}{dt}\bar{\xi}_{2j} = -M_0\,\mathrm{e}^{-2\alpha L}\,\mathrm{e}^{-\alpha(\bar{\xi}_{2j+1}-\bar{\xi}_{2j})}\sin(\omega(\bar{\xi}_{2j+1}-\bar{\xi}_{2j}))+ \\ \qquad\qquad +\mu R(t, L_{2j}+\bar{\xi}_{2j}) + O(\mathrm{e}^{-4(\alpha-\varepsilon)L}), \\ \dfrac{d}{dt}\bar{\xi}_{2j+1} = M_0\,\mathrm{e}^{-2\alpha L}\,\mathrm{e}^{-\alpha(\bar{\xi}_{2j+1}-\bar{\xi}_{2j})}\sin(\omega(\bar{\xi}_{2j+1}-\bar{\xi}_{2j}))+ \\ \qquad\qquad +\mu R(t, L_{2j+1}+\bar{\xi}_{2j+1}) + O(\mathrm{e}^{-4(\alpha-\varepsilon)L}), \end{cases} \tag{11.7}$$

where the self-interaction function $\mathcal{R}(\xi) = -C^{-1}\int_R h(t,x)\partial_x V(x-\xi)\,dx$. We now start to construct a special external force $h(t,x)$. We assume that

$$P_x = 4L, \quad P_t := T_0 \cdot M_0^{-1}\,\mathrm{e}^{2\alpha L} \quad \text{and} \quad h(t,x) = \bar{h}(t\cdot M_0\,\mathrm{e}^{-2\alpha L}, x) \tag{11.8}$$

for some smooth function $\bar{h}$ which is T_0-periodic with respect to t and P_x-periodic with respect to x (the period T_0 is independent of L and will be fixed below). Therefore, all of the self-interaction terms are reduced to the following two functions:

$$R_i(\tau,\bar{\xi}) := -c^{-1}\int_R h(\tau,x)\partial_x V(x-\bar{\xi}-iL), \quad i=0,1, \quad \tau := t\cdot M_0\,\mathrm{e}^{-2\alpha L}. \tag{11.9}$$

On the other hand, we recall that equation (11.1) is invariant with respect to discrete spatial shifts $T^2_{4kL} := \mathcal{T}_{4kL}$, consequently, due to Corollary 8.11, the reduced system (11.7) is also invariant with respect to these spatial shifts and the maps $\mathbb{W}(t,\cdot)$ defined in Theorem 8.5 are also invariant. We now note that the spatial shift on length $4L$ of system (11.7) coincides with the shift of indexes by 2 in the above local coordinates. To be more precise, let

$$S_k : l^\infty \to l^\infty, \quad (S_k\vec{\bar{\xi}})_l := \xi_{2k+l}, \quad k,l \in \mathbb{Z}. \tag{11.10}$$

Then, equations (11.7) will be invariant with respect to this group of shifts. Moreover, according to Corollary 8.10, the map $t \to \mathbb{W}(t,\cdot)$ is P_t-periodic with respect to t and, consequently, the reduced system is also P_t-periodic.

We claim that, in order to prove the theorem, it is sufficient to construct the required hyperbolic set $\mathcal{H}^{tr}_{\rm red}$ for the reduced system (11.7) only. Indeed, according to Corollary 9.11, the associated trajectory set $\mathcal{H}^{tr}_{\rm full}$ of the whole system will be also hyperbolic. The commutation relations (11.4) will be immediate corollaries the analogous relations for the hyperbolic set $\mathcal{H}^{tr}_{\rm red}$. Finally, according to Theorem 8.5 and estimate (8.27), the map $\mathbb{U}_t\ \mathbf{m}(\cdot) \to u(\cdot) := \mathbf{m}(\cdot) + \mathbb{W}(\cdot, \mathbf{m}(\cdot))$ is a homeomorphisms of $\mathcal{H}^{tr}_{\rm red}$ and $\mathcal{H}^{tr}_{\rm full}$ in local topologies as well and, therefore, the continuity of κ is also immediate if $\mathcal{H}^{tr}_{\rm red}$ is homeomorphic to $\mathcal{M}^2$.

Thus, the theorem would be proven if we find a hyperbolic set $\mathcal{H}^{tr}_{red} \in C_{\rm b}(\mathbb{R}, l^\infty)$ of trajectories of (11.7) and a one-to-one map $\bar\kappa : \mathcal{M}^2 \to \mathcal{H}^{tr}_{\rm red}$ satisfying

$$T^1_{lP_t} \circ \bar\kappa = \bar\kappa \circ \mathcal{S}^1_l, \quad S_l \circ \bar\kappa = \bar\kappa \circ \mathcal{S}^2_l \tag{11.11}$$

such that $\bar\kappa$ will be a homeomorphism in a local topologies. Therefore, we can now concentrate ourselves on studying the reduced system (11.7). To this end, we first make the change of variables $t \to \tau$ and fix

$$\mu := M_0\, \mathrm{e}^{-2\alpha L}\,. \tag{11.12}$$

Then, (11.7) reads

$$\begin{cases} \dfrac{d}{d\tau}\bar\xi_{2j} = -\,\mathrm{e}^{-\alpha(\bar\xi_{2j+1}-\bar\xi_{2j})}\sin(\omega(\bar\xi_{2j+1}-\bar\xi_{2j}))+ \\ \qquad\qquad +R_0(\tau,\bar\xi_{2j}) + O(\mathrm{e}^{-2(\alpha-\varepsilon)L}), \\ \dfrac{d}{d\tau}\bar\xi_{2j+1} = \mathrm{e}^{-\alpha(\bar\xi_{2j+1}-\bar\xi_{2j})}\sin(\omega(\bar\xi_{2j+1}-\bar\xi_{2j}))+ \\ \qquad\qquad +R_1(\tau,\bar\xi_{2j+1}) + O(\mathrm{e}^{-2(\alpha-\varepsilon)L}). \end{cases} \tag{11.13}$$

We now need to simplify also the functions $R_i(\tau,\xi)$ introduced in (11.9). To this end, we first introduce a cut-off function $\chi_L \in C_0^\infty(\mathbb{R})$ in such way that $\chi_L(x) \equiv 1$ for $x \in [-L+1, L-1]$ and $\chi_L(x) \equiv 0$ for $x \notin [-L, L]$ and then, for every two given smooth T_0-periodic with respect to τ functions $f_1(\tau,x)$ and $f_2(\tau,x)$ we define the external force h_{f_1,f_2} by

$$h_{f_1,f_2}(\tau,x) := -cf_1(\tau,x)\chi_L(x) - cf_2(\tau,x-2L)\chi_L(x-2L), \tag{11.14}$$

for all $x \in [-2L, 2L]$ and extend it space-periodically for all $x \in \mathbb{R}$. Inserting this formula into (11.9) and taking into the account that $V(x)$ decays exponentially as $x \to \infty$, we arrive at

$$R_i(\tau,\xi) = [f_i * (\partial_x V)](\tau,\xi) + O(\mathrm{e}^{-\alpha L}), \quad i = 0, 1 \tag{11.15}$$

where $(f * g)(\xi) := \int_R f(x)g(x-\xi)\,dx$ is a convolution operator. Thus, (11.13) is now transformed to

$$\begin{cases} \frac{d}{d\tau}\bar\xi_{2j} = -\,\mathrm{e}^{-\alpha(\bar\xi_{2j+1}-\bar\xi_{2j})}\sin(\omega(\bar\xi_{2j+1}-\bar\xi_{2j})) + (f_1 * V')(\tau,\xi_{2j}) + O(\mathrm{e}^{-\alpha L}), \\ \frac{d}{d\tau}\bar\xi_{2j+1} = \mathrm{e}^{-\alpha(\bar\xi_{2j+1}-\bar\xi_{2j})}\sin(\omega(\bar\xi_{2j+1}-\bar\xi_{2j})) + (f_2 * V')(\tau,\bar\xi_{2j+1}) + O(\mathrm{e}^{-\alpha L}) \end{cases} \tag{11.16}$$

where the functions f_1 and f_2 can be chosen arbitrarily. We see that the leading part of the asymptotic expansion of equations (11.13) is now independent of L. Moreover, system (11.16) is weakly coupled perturbation of the following uncoupled system:

$$\begin{cases} \frac{d}{d\tau}\bar\xi_{2j} = -\,\mathrm{e}^{-\alpha(\bar\xi_{2j+1}-\bar\xi_{2j})}\sin(\omega(\bar\xi_{2j+1}-\bar\xi_{2j})) + (f_1 * V')(\tau,\xi_{2j}), \\ \frac{d}{d\tau}\bar\xi_{2j+1} = \mathrm{e}^{-\alpha(\bar\xi_{2j+1}-\bar\xi_{2j})}\sin(\omega(\bar\xi_{2j+1}-\bar\xi_{2j})) + (f_2 * V')(\tau,\bar\xi_{2j+1}) \end{cases} \tag{11.17}$$

which is a $\mathbb{Z}^n$ Cartesian product of the system of two ODEs

$$\begin{cases} \frac{d}{d\tau}Y = -\,\mathrm{e}^{-\alpha(Z-Y)}\sin(\omega(Z-Y)) + (f_1 * V')(\tau, Y), \\ \frac{d}{d\tau}Z = \mathrm{e}^{-\alpha(Z-Y)}\sin(\omega(Z-Y)) + (f_2 * V')(\tau, Z). \end{cases} \tag{11.18}$$

Thus, if this system of two scalar ODEs possesses a hyperbolic set $\mathbb{G}_0$ homeomorphic to the one-dimensional Bernoulli scheme $\mathcal{M}^1$ (and such that the Poincare map of (11.18) (we recall that system (11.18) is T_0-periodic in time) is conjugated to the Bernoulli shift $\mathcal{S}_1$ on $\mathcal{M}^1$, then the whole decoupled system (11.17) possesses a hyperbolic set $\mathbb{G} \sim (\mathbb{G}_0)^{\mathbb{Z}} = \mathcal{M}^2$ and the commutation relations (11.11) (with $P_t = T_0$ due to the scaling of time) will be naturally satisfied. Therefore, due to the stability theorem for hyperbolic sets on $\mathbb{P}(L)$, see Theorem 9.9 and Corollary 9.10, the weakly coupled system (11.16) will also possess a hyperbolic set $\mathcal{H}^{tr}_{\text{red}} \sim \mathcal{M}^2$ which satisfies the same commutation relations.

We also recall that, in order to verify the existence of the above hyperbolic set for system (11.18), it is sufficient to construct at least one transversal (=hyperbolic in the sense of Definition 9.1) homoclinic trajectory for system (11.18), see e.g. [**KaH95**]. Thus, in order to finish the proof of the theorem, we only need the following lemma.

LEMMA 11.3. *There exist $T_0 \in \mathbb{R}$ and smooth T_0-periodic with respect to time functions f_1 and f_2 such that equation* (11.18) *possesses a T_0-periodic in time solution $(Y_p(t), Z_p(t))$ and a transversal homoclinic orbit $(Y_0(t), Z_0(t))$ to it.*

PROOF. We first show that the convolution operators $f_i * V'$ in equations (11.18) can be replaced by arbitrary functions g_i in equations (11.18). To this end, we consider the particular case of function f which is a trigonometric polynomial with respect to x:

$$f(\tau, x) = \sum_{N=-M}^{M} a_N\, \mathrm{e}^{i\kappa N x}, \quad a_N(\tau) \in \mathbb{C} \tag{11.19}$$

when $\kappa > 0$ is a parameter. Then the function $(f * V')(\tau, \xi)$ is also a trigonometric polynomial with respect to ξ and

$$(f * V')(\tau, \xi) = \sum_{N=-M}^{M} b_N(\tau)\, \mathrm{e}^{-i\kappa N\xi}, \quad b_N(\tau) := i\kappa N a_N(\tau)\widehat{V}(\kappa N) \tag{11.20}$$

where $\widehat{V}(\xi)$ is a Fourier transform of the pulse V. We note that the initial pulse is assumed to be exponentially decaying as $x \to \infty$. Consequently, its Fourier transform $\widehat{V}(\xi)$ is an analytic function in a strip $|\operatorname{Im}\xi| < \alpha$ and therefore, it possesses at most countable number of zeros at the real line. Thus, there exists at most countable set $\Xi \subset \mathbb{R}_+$ such that for any $\kappa \in \Xi$, the grid $\{\kappa N,\ N \in \mathbb{Z},\ N \neq 0\}$ contains at least one zero of the function $\widehat{V}(\xi)$. Consequently, for almost all κ, we have

$$\widehat{V}(\kappa N) \neq 0, \quad N \in \mathbb{Z}, \ \ n \neq 0 \tag{11.21}$$

and, therefore, we can invert the expression for $b_N(\tau)$ and find, for any $b_N(\tau)$ with $N \neq 0$ the associated coefficient $a_N(\tau)$. The latter means that, every trigonometric polynomial of the form (11.20) (for almost all κ) with zero mean (i.e. with $b_0 = 0$) can be realized as a convolution of V' with the appropriate function f. Since these polynomials are obviously dense in $C^1(\mathbb{R} \times [-R, R])$ for any $R \in \mathbb{R}$ and the homoclinic orbits preserve under small perturbations, in order to prove the lemma, it is sufficient to construct smooth T_0-periodic with respect to time functions g_1 and g_2 in such way that the following system of two ODEs

$$(11.22)\qquad \begin{cases} \frac{d}{d\tau}Y = -\,\mathrm{e}^{-\alpha(Z-Y)}\sin(\omega(Z-Y)) + g_1(\tau, Y), \\ \frac{d}{d\tau}Z = \mathrm{e}^{-\alpha(Z-Y)}\sin(\omega(Z-Y)) + g_2(\tau, Z) \end{cases}$$

possess a zero solution $Y(\tau) \equiv Z(\tau) \equiv 0$ and a transversal homoclinic orbit (Y_0, Z_0) to that solution.

We are now going to construct "by hands" the homoclinic orbit $(Y_0(t), Z_0(t))$ of problem (11.22). To this end, we first need the origin $Y = Z = 0$ to be a hyperbolic equilibrium. In order to have this property, we will seek for equation (11.22) in the following form:

$$(11.23)\qquad Y' = f_1(Y, Z) + \bar{g}_1(\tau, Y), \quad Z' = f_2(Y, Z) + \bar{g}_2(\tau, Z)$$

where

$$f_1(Y, Z) = -\,\mathrm{e}^{-\alpha(Z-Y)}\sin(\omega(Z-Y)) - 2\omega Y, \quad f_2(Y, Z) := \mathrm{e}^{-\alpha(Z-Y)}\sin(\omega(Z-Y))$$

and the functions $\bar{g}_i(\tau, \xi)$ vanish identically for small $|\xi|$. Obviously, system (11.23) has the form of (11.22). On the other hand, this system has indeed an equilibrium $Y = Z = 0$ and the linearization at this equilibrium reads

$$(11.24)\qquad y' = -\omega(y + z), \quad z' = \omega(z - y).$$

It is not difficult to verify that the matrix of this linear system has the eigenvalues $\lambda_\pm = \pm\omega\sqrt{2}$ and the with the associated eigenvectors $e_\pm = (1, -1\mp\sqrt{2})$. Therefore, the equilibrium is indeed hyperbolic and, moreover, the nonlinear system (11.23) possesses one dimensional stable $\mathcal{V}_-$ and unstable $\mathcal{V}_+$ manifolds near the origin. Since the perturbations $\bar{g}_i$ vanish near the origin, these manifolds are independent of $\bar{g}_i$.

Let us now fix small $\delta > 0$ and construct some special solutions of (11.23). Firstly, let $(Y_\pm(t), Z_\pm(t))$ be the solutions of (11.23) belonging to the manifolds $\mathcal{V}_\pm$. We normalize this solutions by the following condition:

$$(11.25)\qquad Z_+(0) = -\delta, \quad Z_-(0) = \delta.$$

Then we have $\lim_{\tau\pm\infty}(Y_\pm(\tau), Z_\pm(\tau)) = (0, 0)$ and $(Y_\pm(\tau), Z_\pm(\tau))$ are close to the direction of $e_\pm$ for $\pm\tau \geq 0$. We also fix a time $t_0 > 0$ and $y_0 > 0$ such that $\sin\omega y_0 = -1$ and define $Z_n(t)$ as solution of the following equation:

$$(11.26)\qquad Z' = \mathrm{e}^{-\alpha(Z-y_0)}\sin(\omega(Z - y_0)), \quad Z(t_0) = -\delta.$$

Since $\sin\omega y_0 = -1$, then, for sufficiently small δ, the solution $Z_n(t)$ monotonely increases for $t \geq t_0$ and there exists $t_n > 0$ such that $Z_n(t_0 + t_n) = +\delta$.

We are now ready to construct the required homoclinic loop $(Y_0(t), Z_0(t))$ as follows:

$$(11.27)\qquad (Y_0(\tau), Z_0(\tau)) = \begin{cases} (Y_+(\tau), Z_+(\tau)) & \tau \leq 0 \\ (y_0, Z_n(\tau)) & \tau \in [t_0, t_0 + t_n] \\ (Y_-(\tau - t_n - 2t_0), Z_-(\tau - t_n - 2t_0)) & t \geq 2t_0 + t_n. \end{cases}$$

and we connect these pieces of curves in C^∞-smooth way on the time intervals $\tau \in [0, t_0]$ and $\tau \in [t_0 + t_n, 2t_0 + t_n]$ arbitrarily satisfying the only two conditions

$$\begin{aligned} &1)\ Y_0(\tau) \geq Y_0(0),\ Z_0(\tau) \leq -\delta,\quad \tau \in [0, t_0], \\ &2)\ Y_0(\tau) > Y_-(0),\ Z_0(t) \geq \delta,\quad \tau \in [t_0 + t_n, t_0 + 2t_n]. \end{aligned} \tag{11.28}$$

We are now ready to define the perturbations $\bar{g}_i$ in such way that the curve $(Y_0(\tau), Z_0(\tau))$ will be a homoclinic loop for (11.23). To this end, we introduce the cut-off function $\chi_\delta \in C^\infty$ such that $\chi_\delta(x) = 1$ for $|x| < \delta/4$ and $\chi_\delta(x) = 0$ for $|x| \geq \delta/3$ and set

$$\bar{g}_1(\tau, Y) := \begin{cases} 0, \quad \tau \in (-\infty, 0] \cup [2t_0 + t_n, +\infty) \\ \chi_\delta(Y - Y_0(\tau))[Y_0'(\tau) - f_1(Y_0(\tau), Z_0(\tau))], \quad \text{otherwise} \end{cases} \tag{11.29}$$

and

$$\bar{g}_2(\tau, Z) := \begin{cases} 0, \quad \tau \in (-\infty, 0] \cup [t_0, t_0 + t_n] \cup [2t_0 + t_n, +\infty) \\ \chi_\delta(Z - Z_0(\tau))[Z_0'(\tau) - f_2(Y_0(\tau), Z_0(\tau))], \quad \text{otherwise.} \end{cases} \tag{11.30}$$

It then follows form the explicit construction given above that the functions $\bar{g}_i$ thus defined are C^∞-smooth and the function $(Y_0(\tau), Z_0(\tau))$ is indeed a homoclinic loop solution of equation (11.23).

Moreover, the functions $\bar{g}_i$ vanish identically for $\tau \notin [0, 2t_0 + t_n]$ and also in a $\delta/4$-neighborhood of the origin. The latter means that we can extend the functions $\bar{g}_i$ T_0-periodically (with a sufficiently large $T_0 \gg 2t_0 + t_n$) and the homoclinic loop $(Y_0(\tau), Z_0(\tau))$ will be not destroyed.

Thus, the required homoclinic loop $(Y_0(\tau), Z_0(\tau))$ is constructed. In order to finish the proof of the lemma, it only remains to verify that this loop is transversal (hyperbolic). In order to verify that fact, we need to study the linearization of equation (11.23) at the above loop which gives the following operator

$$\mathcal{L} := \frac{\mathrm{d}}{\mathrm{d}\tau} - \begin{pmatrix} \partial_Y[f_1(Y_0, Z_0) + \bar{g}(\tau, Y_0)] & \partial_Z f_1(Y_0, Z_0) \\ \partial_Y f_2(Y_0, Z_0) & \partial_Z[f_2(Y_0, Z_0) + \bar{g}_2(\tau, Z_0)] \end{pmatrix} \tag{11.31}$$

and, in order to have the hyperbolicity, we need this operator to be invertible (say, in $[L^2_b(\mathbb{R})]^2$ or, which is the same, in $[L^2(\mathbb{R})]^2$). We note that, due to the fact that zero equilibrium of (11.23) is hyperbolic and the functions $Y_0(\tau)$ and $Z_0(\tau)$ decay exponentially the essential spectrum of the operator $\mathcal{L}$ (in $[L^2(\mathbb{R})]^2$) coincides with the essential spectrum of the linearization (11.24) which is two imaginary lines $\sigma_{ess}(\mathcal{L}) = \{\pm\omega i\sqrt{2}\lambda,\ \ \lambda \in \mathbb{R}\}$. Thus, if zero belongs to the spectrum of $\mathcal{L}$ (otherwise the operator $\mathcal{L}$ is invertible and, consequently, the homoclinic loop (Y_0, Z_0) is hyperbolic and Lemma 11.3 is proven), it should be only a discrete eigenvalue of finite multiplicity and the associated eigenfunctions (y_0, z_0) decay exponentially as $\tau \to \pm\infty$:

$$\mathcal{L}\begin{pmatrix} y_0 \\ z_0 \end{pmatrix} = 0, \quad |y_0(\tau)| + |z_0(\tau)| \leq c\,\mathrm{e}^{-\omega\sqrt{2}|\tau|}. \tag{11.32}$$

Moreover, since the right-hand side of (11.24) has zero trace, there exists at most one eigenfunction (y_0, z_0) satisfies (11.32) and, consequently, zero eigenvalue of $\mathcal{L}$ has geometric multiplicity one. We introduce also the conjugate eigenfunction (y_0^*, z_0^*) via

$$\mathcal{L}^*\begin{pmatrix} y_0^* \\ z_0^* \end{pmatrix} = 0, \quad |y_0^*(\tau)| + |z_0^*(\tau)| \leq c\,\mathrm{e}^{-\omega\sqrt{2}|\tau|}. \tag{11.33}$$

We are now going to construct a small perturbation of system (11.23) which does not change the loop (Y_0, Z_0), but makes it hyperbolic. To this end, we find positive δ and ν such that

$$Y_0(\tau) \geq \delta, \quad Z_0(\tau) \geq \delta, \quad Y_0'(\tau) \neq Z_0'(\tau) \quad \tau \in [t_c, t_c + \nu] \tag{11.34}$$

for some $t_c \in [0, 2t_0 + t_n]$ (they are exists, due to the explicit construction of the pulse (Y_0, Z_0)) and perturb the functions $\bar{g}_i$ in equations (11.23) as follows:

$$\begin{aligned} \tilde{g}_1(\tau, Y) &:= \bar{g}_1(\tau, Y) + \varepsilon C_1(\tau)\chi_\delta(Y - Y_0(\tau))(Y - Y_0(\tau)), \\ \tilde{g}_2(\tau, Z) &:= \bar{g}_2(\tau, Z) + \varepsilon C_2(\tau)\chi_\delta(Z - Z_0(\tau))(Z - Z_0(\tau)), \end{aligned} \tag{11.35}$$

for $t \in [0, 2t_0 + t_n]$ and then extend them T_0-periodically for all τ. Here $\varepsilon > 0$ is a small parameter and C_i are smooth functions satisfying

$$\operatorname{supp} C_i \subset [t_c, t_c + \nu]. \tag{11.36}$$

Then, obviously the perturbation of the form (11.35) does not destroy the homoclinic loop (Y_0, Z_0) (it solves the perturbed equations as well) and changes the linearization operator (11.31) as follows:

$$\tilde{\mathcal{L}}_\varepsilon = \mathcal{L} - \varepsilon \begin{pmatrix} C_1(\tau) & 0 \\ 0 & C_2(\tau) \end{pmatrix} \tag{11.37}$$

where the functions C_i satisfying (11.36) and $\varepsilon > 0$ can be chosen arbitrarily. We claim that it is possible to chose these parameters in such way that $0 \notin \sigma(\tilde{\mathcal{L}}_\varepsilon)$. Then, the loop (Y_0, Z_0) would be hyperbolic for the perturbed equation and the lemma would be proven. Assume that the latter is not true. Then, for any $\varepsilon \in \mathbb{R}$ and any functions C_i satisfying (11.36), we have

$$0 \in \sigma(\tilde{\mathcal{L}}_\varepsilon). \tag{11.38}$$

Then, exactly as for the case of non-perturbed equation, there exists an exponentially decaying eigenfunction $(y_\varepsilon, z_\varepsilon)$ of the perturbed operator (11.37):

$$\tilde{\mathcal{L}}_\varepsilon \begin{pmatrix} y_\varepsilon \\ z_\varepsilon \end{pmatrix} = 0, \quad |y_\varepsilon(\tau)| + |z_\varepsilon(\tau)| \leq c\, e^{-\omega\sqrt{2}|\tau|} \tag{11.39}$$

whose geometrical multiplicity is one. Moreover, due to the classical spectral perturbation theory, the functions $(y_\varepsilon, z_\varepsilon)$ tend to (y_0, z_0) as $\varepsilon \to 0$, e.g. in $C_b(\mathbb{R})$ (under the proper normalization). On the other hand, multiplying equation (11.39) scalarly in $[L^2(\mathbb{R})]^2$ by the function (y_0^*, z_0^*) and using (11.33), we infer

$$\int_{t_c}^{t_c+\nu} [C_1(\tau)y_\varepsilon(\tau)y_0^*(\tau) + C_2(\tau)z_\varepsilon(\tau)z_0^*(\tau)]\, d\tau = 0. \tag{11.40}$$

Passing now to the limit $\varepsilon \to 0$ in (11.40), we finally arrive at

$$\int_{t_c}^{t_c+\nu} [C_1(\tau)y_0(\tau)y_0^*(\tau) + C_2(\tau)z_0(\tau)z_0^*(\tau)]\, d\tau = 0 \tag{11.41}$$

which should be valid for any C_i satisfying (11.36). The latter immediately implies that

$$y_0(\tau) \cdot y_0^*(\tau) \equiv z_0(\tau) \cdot z_0^*(\tau) \equiv 0, \quad \tau \in [t_c, t_c + \nu].$$

Thus, without loss of generality, we can conclude that $y_0(\tau) \equiv 0$ on some subinterval $J \subset [t_c, t_c+\nu]$ (otherwise, the same assertion for $z_0(\tau)$ or for conjugate eigenfunction is true). We also recall that, due to the assumption $Y_0(\tau) \neq Z_0(\tau)$ for $\tau \in [t_c, t_c+\nu]$, we obtain that the functions

$$\partial_Z f_1((Y_0(\tau), Z_0(\tau)) \quad \text{and} \quad \partial_Y f_2(Y_0(\tau), Z_0(\tau))$$

do not vanish identically on $[t_c, t_c + \nu]$. Thus, from equation (11.32) we conclude that $z_0(\tau) \equiv 0$ on J as well which is impossible since (y_0, z_0) is a nontrivial solutions of linear system of two ODEs. Thus, our assumption is wrong and there exist ε and C_i such that (11.38) is wrong. Thus, Lemma 11.3 is proven. □

Hence, theorem 11.2 is proven. □

REMARK 11.4. We see that equation (11.1) with the spatio-temporal external forcing of an arbitrarily small order μ possesses a hyperbolic set homeomorphic to $\mathcal{M}^2$. Nevertheless, in the case of Swift-Hohenberg equation, we clearly cannot take $\mu = 0$. Indeed, in this case the equations of two-pulse interactions have a gradient structure (and the same is true for N-pulse interaction for any finite N) and autonomous and, consequently, cannot generate the chaotic in time oscillations. The same problem arises if we try to restrict ourselves by considering only time independent external forces.

Moreover, equation (11.1) is formally gradient, consequently, the spatio-temporal topological entropy of any spatio-temporal invariant subset of it (e.g. global attractor if it exists) equals zero, see [**Zel04**]. Since this entropy is strictly positive for the Bernoulli scheme $\mathcal{M}^2$, the homeomorphic embedding of it into the space of trajectories of (11.1) satisfying the commutation relations (11.4) cannot exist in the case $\mu = 0$.

In contrast to this, for the case of generalized Ginzburg-Landau equation considered in Example 10.8, the two-pulse interaction system is far from being gradient and can (in principle) generate chaotic dynamics in time. Thus, it is probably possible to find the space-time chaotic dynamics in the Ginzburg-Landau equations even without external forces (or for small *autonomous* space periodic external forces). We return to this problem somewhere else.

REMARK 11.5. As we have already mentioned before, the simplest nonlinearity f for which the non-perturbed Swift-Hohenberg equation possesses a pulse to zero equilibrium has the form $f(u) = u^3 + \kappa u^2$ (for some values of parameters κ and β) which differs from the classical nonlinearity $f(u) = u^3$ by presence of a sufficiently large quadratic term κu^2 (it is also known that the cubic Swift-Hohenberg equation with $\kappa = 0$ does not possess equilibria *homoclinic to zero state, see* [**PeT01**]. However, the term κu^2 can be vanished by the trivial linear variable change $u = \bar{u} - \kappa/3$ in equation (11.1) which gives

$$\partial_t \bar{u} + (\partial_x^2 + 1)^2 \bar{u} + \bar{u}^3 - \beta' \bar{u} = \kappa' + \mu h(t) := \bar{h}(t) \tag{11.42}$$

for some new values β' and κ'. Thus, the result on existence of Sinai-Bunimovich space-time chaos stated in Theorem 11.2 remains true for the classical cubic nonlinearity as well (only the external forces should contain in that case the additional non-small constant κ'.

To conclude, we note that we have proved Theorem 11.2 by constructing highly artificial space-time periodic external forces $h(t, x)$. This were made in order to avoid rather delicate dynamical analysis of the two-pulse interaction equations (11.18) and to have the *explicit* expression for the transversal homoclinic orbit in it.

On the other hand, everything what we need from this system is the existence of the irregular recurrent dynamics ($\sim$ the existence of a single transversal homoclinic orbit) which does not look as a great restriction. Thus, the result of Theorem 11.2 is expected to be true for large class of space-time periodic external forces $h(t, x)$. The only essential restrictions on this class are the following ones:

1) Sufficiently large spatial period P_x (in order to be able to have different influence to different elements of the pulse-pair).

2) Essentially larger time period $P_t \sim e^{\beta P_x}$. Indeed, the characteristic rate of pulse evolution is proportional to $e^{-2\alpha L}$ and if the time period will be essentially shorter the influence of the non-autonomous effects will be averaged and we return to the autonomous gradient system which cannot demonstrate chaos.

Up to these structural assumptions (which seem to be unavoidable under the above method), the external forces $h(t, x)$ are expected to be more-or less arbitrary.

Bibliography

[ABC96] V. AFRAIMOVICH, A. V. BABIN AND S. N. CHOW: Spatial chaotic structure of attractors of reaction-diffusion systems, *Trans. Amer. Math. Soc.*, Vol. 348, 1996, no. 12, 5031–5063.

[AfM99] A. AFENDIKOV AND A. MIELKE: Bifurcation of homoclinic orbits to a saddle-focus in reversible systems with SO(2)-symmetry, *J. Diff. Eqns* Vol. 159, 1999, no. 2, 370–402.

[AfM01] A. AFENDIKOV and A. MIELKE: Multi-pulse solutions to the Navier-Stokes problem between parallel plates. *Z. Angew. Math. Phys.*, Vol. 52(1), 79–100, 2001.

[AfF00] V. AFRAIMOVICH and B. FERNANDEZ: Topological properties of linearly coupled map lattices, *Nonlinearity*, Vol. 13, 2000, 973–993

[AgN63] S. AGMON, L. NIRENBERG: Properties of solutions of ordinary differential equations in Banach space. *Comm. Pure Appl. Math*, Vol. 16, 1963, 121–239.

[AgN67] S. AGMON, L. NIRENBERG: Lower bounds and uniqueness theorems for solutions of differential equations in a Hilbert space, *Comm. Pure Appl. Math.*, Vol. 20, 1967, 207–229.

[ASCT01] N. AKHMEDIEV, J. SOTO-CRESPO AND G. TOWN: Pulsating Solitons, chaotic solitons, period doubling, and pulse coexistence in mode-locked lasers: Complex Ginzburg-Landau Equation Approach. *Phys. Rev. E*, Vol. 63, 2001, 056602-1–056602-13.

[Ama95] H. AMANN: *Linear and quasilinear parabolic problems. Vol. I: Abstract linear theory.* Monographs in Mathematics Vol. 89, Birkhauser Boston 1995.

[Ang87] S. ANGENENT: The shadowing lemma for elliptic PDE. In *Dynamics of infinite-dimensional systems (Lisbon, 1986)*, volume 37 of *NATO Adv. Sci. Inst. Ser. F Comput. Systems Sci.*, pages 7–22. Springer, Berlin, 1987.

[Bab00] A.V. BABIN: Topological invariants and solutions with a high complexity for scalar semi-linear PDE, *J. Dynam. Differential Equations*, Vol. 12, 2000, no. 3, 599–646.

[BaS02] P. BATES AND J. SHI: Existence and Instability of Spike Layer Solutions to Singular Perturbation Problems, *J. Functional Analysis*, Vol. 196, 2002, 211-264.

[BeL83] H. BERESTYCKI AND P.-L. LIONS: Nonlinear scalar field equations. I. Existence of a ground state. *Arch. Rational Mech. Anal.* 82, 1983, No. 4, 313–345.

[BGL97] L. BELYAKOV, L. GLEBSKY AND L. LERMAN: Abundance of stable stationary localized solutions to the generalized 1D Swift-Hohenberg equation, *Comp. Math. Appl*, Vol. 34, No. 3-4, 1997, 253–266.

[BlW02] S. BLANCHFLOWER AND N. WEISS: Three-dimensional magnetohydrodynamic convectons. *Phys. Lett. A*, Vol.294 (2002), No. 5-6, 297–303.

[BuS88] L.A. BUNIMOVICH AND YA. G. SINAĬ: Spacetime chaos in coupled map lattices, *Nonlinearity*, Vol. 1, 1988, no. 4, 491–516.

[CoE99a] P. COLLET AND J.-P. ECKMANN: The definition and measurement of the topological entropy per unit volume in parabolic PDEs, *Nonlinearity* Vol. 12, 1999, no. 3, 451–473.

[CoM03] S. COX AND P. MATTHEWS: Instability and localization of patterns due to a conserved quantity. *Phys. D*, Vol. 175, 2003, no. 3-4, 196–219.

[Cou02] P. COULLET: Localized patterns and fronts in nonequilibrium systems, *Int. J of Bifurcation and Chaos*, Vol. 12, No. 11, 2002, 2445-2457.

[DFKM96] G. DANGELMAYR, B. FIEDLER, K. KIRCHGÄSSNER, and A. MIELKE: *Dynamics of nonlinear waves in dissipative systems: reduction, bifurcation and stability*, volume 352 of *Pitman Research Notes in Mathematics Series*. Longman, Harlow, 1996. With a contribution by G. Raugel.

[DGK98] A. DOELMANN, A. GARDNER AND T. KAPER: Stability analysis of singular patterns in the 1D Gray-Scott model, *Physica D*, 122, 1–36.

[EcR98] J.-P. ECKMANN AND J. ROUGEMONT: Coarsening by Ginzburg-Landau dynamics. *Comm. Math. Phys. 199*, No. 2, 1998, 441–470.

[Ei2002] S-I. EI: The motion of weakly interacting pulses in reaction-diffusion systems, *JDDE*, Vol. 14, No. 1, 2002, 85–137.

[EfZ01] M. EFENDIEV, S. ZELIK: The attractor for a nonlinear reaction–diffusion system in an unbounded domain, *Comm. Pure Appl. Math.* Vol. 54, 2001, 625–688.

[GlL94] L. GLEBSKY AND L. LERMAN: On small stationary localized solutions for the generalized 1D Swift-Hohenberg equation, *Chaos*, 5(2), 1994, 424–431.

[GEP98] G. GOREN, J.-P. ECKMANN, and I. PROCACCIA: Scenario for the onset of space-time chaos. *Phys. Rev. E (3)*, 57(4), 4106–4134, 1998.

[Hen81] D. HENRY: Geometric theory of semilinear parabolic equations. *Lecture Notes in Mathematics*, 840. Springer-Verlag, Berlin-New York, 1981.

[KaH95] A. KATOK, B. HASSELBLATT: *Introduction to the Modern Theory of Dynamical Systems*, Cambridge University Press, 1995.

[Kir82] K. KIRCHGÄSSNER: Wave-solutions of reversible systems and applications. *J. Differential Equations*, 45(1), 113–127, 1982.

[Kir85] K. KIRCHGÄSSNER: Nonlinear wave motion and homoclinic bifurcation. In *Theoretical and Applied Mechanics (Lyngby, 1984)*, pages 219–231. North-Holland, Amsterdam, 1985.

[LSU67] O.A. LADYZHENSKAYA, V.A. SOLONNIKOV, N.N. URALTSEVA: *Linear and Quasilinear Equations of Parabolic Type*, Nauka, 1967.

[Man90] P. MANNEVILLE: *Dissipative structures and weak turbulence.* Perspectives in Physics. Academic Press Inc., Boston, MA, 1990.

[Man95] P. MANNEVILLE: Dynamical systems, temporal vs. spatio-temporal chaos, and climate. In *Nonlinear Dynamics and Pattern Formation in the Natural Environment (Noordwijkerhout, 1994)*, volume 335 of *Pitman Res. Notes Math. Ser.*, pages 168–187. Longman, Harlow, 1995.

[Mie86] A. MIELKE: A reduction principle for nonautonomous systems in infinite-dimensional spaces. *J. Differential Equations*, 65(1), 68–88, 1986.

[Mie88] A. MIELKE: Reduction of quasilinear elliptic equations in cylindrical domains with applications. *Math. Meth. Appl. Sci.*, 10, 1988, 51–66.

[Mie90] A.MIELKE: Normal hyperbolicity of center manifolds and Saint-Venant's principle. *Arch. Rat. Mech Anal.*, 110, 1990, 353–372.

[Mie97] A. MIELKE: Instability and stability of rolls in the Swift-Hohenberg equation, *Comm. Math. Phys.* Vol. 189, 1997, 829–853.

[Mie02] A. MIELKE: The Ginzburg–Landau equation in its role as a modulation equation. In "Handbook of Dynamical Systems, Vol. 2, B. Fiedler (edr), Elsevier 2002," pp. 759–834.

[MiH88] A. MIELKE and P. HOLMES: Spatially complex equilibria of buckled rods. *Arch. Rational Mech. Anal.*, 101(4), 319–348, 1988.

[MiZ04] A.MIELKE AND S. ZELIK: Infinite-dimensional hyperbolic sets and spatio-temporal chaos in reaction-diffusion systems in $\mathbb{R}^n$. Jour. Dyn. Diff. Eqns., Vol. 19, issue 2, 2007, 333–389.

[PeT01] L. PELETIER AND W.TROY: Spatial patterns. Higher order models in physics and mechanics. *Progress in Nonlinear Differential Equations and their Applications*, 45. Birkhï¿$\frac{1}{2}$ser Boston, Inc., Boston, MA, 2001.

[PeS88] Y.B. PESIN and Y.G. SINAĬ: Space-time chaos in the system of weakly interacting hyperbolic systems. *J. Geom. Phys.*, 5(3), 483–492, 1988.

[PeS91] Y.B. PESIN AND Y.G. SINAĬ: Space-time chaos in chains of weakly interacting hyperbolic mappings,*Adv. Soviet Math., 3, Dynamical systems and statistical mechanics (Moscow, 1991)*, 165–198, Amer. Math. Soc., Providence, RI, 1991.

[Rab93] P.H. RABINOWITZ: Multibump solutions of a semilinear elliptic PDE on $\mathbb{R}^n$, *Degenerate diffusions (Minneapolis, MN, 1991)*, 175–185, *IMA Vol. Math. Appl.*, 47, Springer, New York, 1993.

[REW00] M. I. RABINOVICH, A. B. EZERSKY AND P. D. WEIDMAN: The dynamics of patterns. *World Scientific Publishing Co.*, Inc., River Edge, NJ, 2000.

[Sak90] K. SAKAMOTO: Invariant manifolds in singular perturbation problems for ordinary differential equations *Proc. Roy. Soc. Edinburgh A*, 116, 1990, 45–78.

[San93] B. SANDSTEDE: Verzweigungtheorie homokliner Verdopplungen, *PhD thesis*, University of Stuttgart, 1993.

[San02] B. Sandstede: Stability of traveling waves. *Handbook of dynamical systems*, Vol. 2, 983–1055, North-Holland, Amsterdam, 2002.

[Sand] B. Sandstede: Weak interaction of pulses (in preparation)

[SSTC01] L. P. Shilnikov, A. L. Shilnikov, D. Turaev and L. Chua: Methods of qualitative theory in nonlinear dynamics. Part II, *World Scientific Series on Nonlinear Science. Series A: Monographs and Treatises, 5. World Scientific Publishing Co., Inc., River Edge, NJ*, 2001.

[SkV02] D. Skryabin and A. Vladimirov: Vortex induced rotation of clusters of localized states in the complex Ginzburg-Landau equation. *Phys. Rev. Let.* Vol. 89, No. 4, 2002, 044101-1–044101-4.

[TVM03] M. Tlidi, A. Vladimirov and P. Mandel: Interaction and stability of periodic and localized structures in optical bistable systems. *IEEE Jour. of Quantum Electronics*, Vol. 39, No. 2, 2003, 216–226.

[Tri78] H. Triebel: *Interpolation Theory, Function Spaces, Differential Operators*, North–Holland, 1978.

[TVZ07] D. Turaev, A. Vladimirov and S.Zelik: Chaotic bound state of localized structures in the complex Ginzburg-Landau equation, *Phys. Rev. E*, Vol. 75, 2007, pp. 045601(R).

[TZ07a] D. Turaev and S. Zelik: Infinite-dimensional hyperbolic attractors for space-time periodically perturbed 1D Swift-Hohenberg equations, 2007 (in preparation).

[TZ07b] D. Turaev and S. Zelik: Sinai-Bunimovich space-time chaos in 1D Ginzburg-Landau equations: the homogeneous case, 2007 (in preparation).

[VaV87] S. van Gils and A. Vanderbauwhede: Center manifolds and contractions on a scale of Banach spaces. *J. Funct. Anal.*, Vol. 72, 1987, 209–224.

[VFK*99] A. Vladimirov, S. Fedorov, N. Kaliteevskii, G. Khodova and N. Rosanov: Numerical investigation of laser localized structures. *J. Opt. B: Quantum Semicl. Opt.*, Vol. 1, 1999, 101–106.

[VKR01] A. Vladimirov, G. Khodova and N. Rosanov: Stable bound states of one-dimensional autosolitons in a bistable laser. *Phys. Rev. Let*, Vol. 63, 2001, 056607-1–056607-6.

[VMSF02] A. Vladimirov, J. McSloy, D.Skryabin and W. Firth: Two-dimensional clusters of solitary structures in driven optical cavities. *Phys. Rev. E*, Vol. 65, 2002, 046606-1–046606-11.

[Zel00] S.V. Zelik: Spatial and dynamical chaos generated by reaction diffusion systems in unbounded domains, *J. Dynam. Diff. Eqns.*, Vol. 19, issue 1, 2007, 1–74.

[Zel03a] S.V. Zelik: The attractor for an nonlinear reaction–diffusion system in an unbounded domain and Kolmogorov's ϵ–entropy, *Mathem. Nachr.*, Vol. 248-249, issue 1, 2003, 72–96.

[Zel03b] S.V. Zelik: The attractors of reaction-diffusion systems in unbounded domains and their spatial complexity, *Comm. Pure Appl. Math.*, Vol. 56, issue 5, 2003, 584–637.

[Zel04] S. V. Zelik: Multiparametrical semigroups and attractors of reaction- diffusion systems in $\mathbb{R}^n$, *Proc. Moscow Math. Soc.*, Vol. 65, 2004, 69–130.

Nomenclature

$2L$ – minimal admissible distance between pulses, page 25
$\alpha \to \Gamma(\alpha)$ – local coordinates on G, page 13
$\mathbb{B}(L)$ – space of admissible pulse configurations, page 25
$\mathbb{P}(L)$ – multi-pulse manifold, page 27
$\mathbb{P}_1$ – one-pulse manifold, page 12
$\mathbb{P}_\Gamma$ – spectral projector on the kernel of $\mathcal{L}_\Gamma$, page 19
$\mathbb{P}_\Gamma^*$ – conjugate spectral projector, page 19
$\mathbb{P}_\mathbf{m}$ – linear projector to the tangent space of $\mathbb{P}(L)$ at $\mathbf{m}$, page 35
$\mathbb{W} = \mathbb{W}(t, \mathbf{m})$ – center manifold reduction map, page 58
$\mathcal{H}^{tr}$ – hyperbolic set of trajectories, page 68
$\mathcal{H}^{tr}_{\text{full}}$ – hyperbolic set of the initial system, page 73
$\mathcal{H}^{tr}_{\text{red}}$ – reduced hyperbolic set on the center manifold, page 73
$\mathcal{L}_\Gamma$ – linearization on the shifted pulse V_Γ, page 12
$\mathcal{L}_V$ – linearization on the initial pulse V, page 11
$\mathcal{T}_\Gamma$ – action of Γ on the space of functions, page 9
$\Gamma \in G$ – element of the Lie group G, page 9
$\mathbf{m}$ – element of $\mathbb{P}(L)$
$\mathrm{St}_G(V)$ – isotropy group of V, page 12
$\partial\mathbb{P}(L)$ – boundary of the manifold $\mathbb{P}(L)$, page 30
$\Phi(u)$ – nonlinearity, page 8
ϕ^i – eigenvectors of $\mathcal{L}_V$, page 11
ϕ^i_Γ – shifted eigenfunctions, page 13
$\pi(v)$ – nonlinear projector to the manifold $\mathbb{P}(L)$, page 45
$\Pi_{ij}(\alpha)$ – transfer matrix, page 13
ψ^i – conjugate eigenvectors for $\mathcal{L}_V$, page 11
ψ^i_Γ – shifted conjugate eigenfunctions, page 13
$\mathrm{sep}(\vec{\Gamma})$ – minimal distance between pulses, page 25
$\theta = \theta(x)$ – weights of exponential growth rate, page 16
$\vec{\Gamma} = \{\Gamma_j\}_{j=1}^{\infty}$ – admissible multi-pulse configuration, page 25
$\{T_\xi, \xi \in \mathbb{R}^n\}$ – group of spatial translations, page 9
A_0 – linear elliptic operator of order $2l$, page 8
$d^s(\vec{\Gamma}^1, \vec{\Gamma}^2)$ – uniform metric on $\mathbb{B}(L)$, page 25
$d^s_\gamma(\vec{\Gamma}^1, \vec{\Gamma}^2)$ – weighted metric on $\mathbb{B}(L)$, page 26
G – Lie group of symmetries, page 9
m – number of equations, page 8
n – spatial dimension, page 8
$T_\mathbf{m}\mathbb{P}(L)$ – tangent space to $\mathbb{P}(L)$ at $\mathbf{m}$, page 30
V – the initial pulse, page 10
V_Γ – shifted pulse, page 12
$W^{2l,p}(\mathbb{R}^n)$ – usual Sobolev spaces, page 9
$W^{l,p}_\theta(\mathbb{R}^n)$ – weighted Sobolev spaces, page 17
$W^{l,p}_{b,\theta}(\mathbb{R}^n)$ – weighted uniformly local spaces, page 17

Editorial Information

To be published in the *Memoirs*, a paper must be correct, new, nontrivial, and significant. Further, it must be well written and of interest to a substantial number of mathematicians. Piecemeal results, such as an inconclusive step toward an unproved major theorem or a minor variation on a known result, are in general not acceptable for publication.

Papers appearing in *Memoirs* are generally at least 80 and not more than 200 published pages in length. Papers less than 80 or more than 200 published pages require the approval of the Managing Editor of the Transactions/Memoirs Editorial Board.

As of November 30, 2008, the backlog for this journal was approximately 11 volumes. This estimate is the result of dividing the number of manuscripts for this journal in the Providence office that have not yet gone to the printer on the above date by the average number of monographs per volume over the previous twelve months, reduced by the number of volumes published in four months (the time necessary for preparing a volume for the printer). (There are 6 volumes per year, each usually containing at least 4 numbers.)

A Consent to Publish and Copyright Agreement is required before a paper will be published in the *Memoirs*. After a paper is accepted for publication, the Providence office will send a Consent to Publish and Copyright Agreement to all authors of the paper. By submitting a paper to the *Memoirs*, authors certify that the results have not been submitted to nor are they under consideration for publication by another journal, conference proceedings, or similar publication.

Information for Authors

Memoirs are printed from camera copy fully prepared by the author. This means that the finished book will look exactly like the copy submitted.

Initial submission. The AMS uses Centralized Manuscript Processing for initial submissions. Authors should submit a PDF file using the Initial Manuscript Submission form found at `www.ams.org/peer-review-submission`, or send one copy of the manuscript to the following address: Centralized Manuscript Processing, MEMOIRS OF THE AMS, 201 Charles Street, Providence, RI 02904-2294 USA. If a paper copy is being forwarded to the AMS, indicate that it is for it Memoirs and include the name of the corresponding author, contact information such as email address or mailing address, and the name of an appropriate Editor to review the paper (see the list of Editors below).

The paper must contain a *descriptive title* and an *abstract* that summarizes the article in language suitable for workers in the general field (algebra, analysis, etc.). The *descriptive title* should be short, but informative; useless or vague phrases such as "some remarks about" or "concerning" should be avoided. The *abstract* should be at least one complete sentence, and at most 300 words. Included with the footnotes to the paper should be the 2000 *Mathematics Subject Classification* representing the primary and secondary subjects of the article. The classifications are accessible from `www.ams.org/msc/`. The list of classifications is also available in print starting with the 1999 annual index of *Mathematical Reviews*. The Mathematics Subject Classification footnote may be followed by a list of *key words and phrases* describing the subject matter of the article and taken from it. Journal abbreviations used in bibliographies are listed in the latest *Mathematical Reviews* annual index. The series abbreviations are also accessible from `www.ams.org/msnhtml/serials.pdf`. To help in preparing and verifying references, the AMS offers MR Lookup, a Reference Tool for Linking, at `www.ams.org/mrlookup/`.

Electronically prepared manuscripts. The AMS encourages electronically prepared manuscripts, with a strong preference for $\mathcal{AMS}$-LaTeX. To this end, the Society has prepared $\mathcal{AMS}$-LaTeX author packages for each AMS publication. Author packages include instructions for preparing electronic manuscripts, samples, and a style file that generates the particular design specifications of that publication series. Though $\mathcal{AMS}$-LaTeX is the highly preferred format of TeX, author packages are also available in $\mathcal{AMS}$-TeX.

Authors may retrieve an author package for *Memoirs of the AMS* from `www.ams.org/journals/memo/memoauthorpac.html` or via FTP to `ftp.ams.org` (login as `anonymous`, enter username as password, and type `cd pub/author-info`). The *AMS Author Handbook* and the *Instruction Manual* are available in PDF format from the author package link. The author package can also be obtained free of charge by sending email to `tech-support@ams.org` (Internet) or from the Publication Division, American Mathematical Society, 201 Charles St., Providence, RI 02904-2294, USA. When requesting an author package, please specify $\mathcal{AMS}$-LaTeX or $\mathcal{AMS}$-TeX and the publication in which your paper will appear. Please be sure to include your complete mailing address.

After acceptance. The final version of the electronic file should be sent to the Providence office (this includes any TeX source file, any graphics files, and the DVI or PostScript file) immediately after the paper has been accepted for publication.

Before sending the source file, be sure you have proofread your paper carefully. The files you send must be the EXACT files used to generate the proof copy that was accepted for publication. For all publications, authors are required to send a printed copy of their paper, which exactly matches the copy approved for publication, along with any graphics that will appear in the paper.

Accepted electronically prepared files can be submitted via the web at `www.ams.org/submit-book-journal/`, sent via FTP, or sent on CD-Rom or diskette to the Electronic Prepress Department, American Mathematical Society, 201 Charles Street, Providence, RI 02904-2294 USA. TeX source files, DVI files, and PostScript files can be transferred over the Internet by FTP to the Internet node `ftp.ams.org` (130.44.1.100). When sending a manuscript electronically via CD-Rom or diskette, please be sure to include a message identifying the paper as a Memoir.

Electronically prepared manuscripts can also be sent via email to `pub-submit@ams.org` (Internet). In order to send files via email, they must be encoded properly. (DVI files are binary and PostScript files tend to be very large.)

Electronic graphics. Comprehensive instructions on preparing graphics are available at `www.ams.org/authors/journals.html`. A few of the major requirements are given here.

Submit files for graphics as EPS (Encapsulated PostScript) files. This includes graphics originated via a graphics application as well as scanned photographs or other computer-generated images. If this is not possible, TIFF files are acceptable as long as they can be opened in Adobe Photoshop or Illustrator. No matter what method was used to produce the graphic, it is necessary to provide a paper copy to the AMS.

Authors using graphics packages for the creation of electronic art should also avoid the use of any lines thinner than 0.5 points in width. Many graphics packages allow the user to specify a "hairline" for a very thin line. Hairlines often look acceptable when proofed on a typical laser printer. However, when produced on a high-resolution laser imagesetter, hairlines become nearly invisible and will be lost entirely in the final printing process.

Screens should be set to values between 15% and 85%. Screens which fall outside of this range are too light or too dark to print correctly. Variations of screens within a graphic should be no less than 10%.

Inquiries. Any inquiries concerning a paper that has been accepted for publication should be sent to `memo-query@ams.org` or directly to the Electronic Prepress Department, American Mathematical Society, 201 Charles St., Providence, RI 02904-2294 USA.

Editors

This journal is designed particularly for long research papers, normally at least 80 pages in length, and groups of cognate papers in pure and applied mathematics. Papers intended for publication in the *Memoirs* should be addressed to one of the following editors. The AMS uses Centralized Manuscript Processing for initial submissions to AMS journals. Authors should follow instructions listed on the Initial Submission page found at www.ams.org/memo/memosubmit.html.

Algebra to ALEXANDER KLESHCHEV, Department of Mathematics, University of Oregon, Eugene, OR 97403-1222; email: ams@noether.uoregon.edu

Algebraic geometry to DAN ABRAMOVICH, Department of Mathematics, Brown University, Box 1917, Providence, RI 02912; email: amsedit@math.brown.edu

Algebraic geometry and its application to MINA TEICHER, Emmy Noether Research Institute for Mathematics, Bar-Ilan University, Ramat-Gan 52900, Israel; email: teicher@macs.biu.ac.il

Algebraic topology to ALEJANDRO ADEM, Department of Mathematics, University of British Columbia, Room 121, 1984 Mathematics Road, Vancouver, British Columbia, Canada V6T 1Z2; email: adem@math.ubc.ca

Combinatorics to JOHN R. STEMBRIDGE, Department of Mathematics, University of Michigan, Ann Arbor, Michigan 48109-1109; email: FRS@umich.edu

Commutative and homological algebra to LUCHEZAR L. AVRAMOV, Department of Mathematics, University of Nebraska, Lincoln, NE 68588-0130; email: avramov@math.unl.edu

Complex analysis and harmonic analysis to ALEXANDER NAGEL, Department of Mathematics, University of Wisconsin, 480 Lincoln Drive, Madison, WI 53706-1313; email: nagel@math.wisc.edu

Differential geometry and global analysis to LISA C. JEFFREY, Department of Mathematics, University of Toronto, 100 St. George St., Toronto, ON Canada M5S 3G3; email: jeffrey@math.toronto.edu

Dynamical systems and ergodic theory and complex anaysis to YUNPING JIANG, Department of Mathematics, CUNY Queens College and Graduate Center, 65-30 Kissena Blvd., Flushing, NY 11367; email: Yunping.Jiang@qc.cuny.edu

Functional analysis and operator algebras to DIMITRI SHLYAKHTENKO, Department of Mathematics, University of California, Los Angeles, CA 90095; email: shlyakht@math.ucla.edu

Geometric analysis to WILLIAM P. MINICOZZI II, Department of Mathematics, Johns Hopkins University, 3400 N. Charles St., Baltimore, MD 21218; email: trans@math.jhu.edu

Geometric analysis to MARK FEIGHN, Math Department, Rutgers University, Newark, NJ 07102; email: feighn@andromeda.rutgers.edu

Harmonic analysis, representation theory, and Lie theory to ROBERT J. STANTON, Department of Mathematics, The Ohio State University, 231 West 18th Avenue, Columbus, OH 43210-1174; email: stanton@math.ohio-state.edu

Logic to STEFFEN LEMPP, Department of Mathematics, University of Wisconsin, 480 Lincoln Drive, Madison, Wisconsin 53706-1388; email: lempp@math.wisc.edu

Number theory to JONATHAN ROGAWSKI, Department of Mathematics, University of California, Los Angeles, CA 90095; email: jonr@math.ucla.edu

Number theory to SHANKAR SEN, Department of Mathematics, 505 Malott Hall, Cornell University, Ithaca, NY 14853; email: ss70@cornell.edu

Partial differential equations to GUSTAVO PONCE, Department of Mathematics, South Hall, Room 6607, University of California, Santa Barbara, CA 93106; email: ponce@math.ucsb.edu

Partial differential equations and dynamical systems to PETER POLACIK, School of Mathematics, University of Minnesota, Minneapolis, MN 55455; email: polacik@math.umn.edu

Probability and statistics to RICHARD BASS, Department of Mathematics, University of Connecticut, Storrs, CT 06269-3009; email: bass@math.uconn.edu

Real analysis and partial differential equations to DANIEL TATARU, Department of Mathematics, University of California, Berkeley, Berkeley, CA 94720; email: tataru@math.berkeley.edu

All other communications to the editors should be addressed to the Managing Editor, ROBERT GURALNICK, Department of Mathematics, University of Southern California, Los Angeles, CA 90089-1113; email: guralnic@math.usc.edu.

Titles in This Series

929 **Richard F. Bass, Xia Chen, and Jay Rosen,** Moderate deviations for the range of planar random walks, 2009

928 **Ulrich Bunke,** Index theory, eta forms, and Deligne cohomology, 2009

927 **N. Chernov and D. Dolgopyat,** Brownian Brownian motion-I, 2009

926 **Riccardo Benedetti and Francesco Bonsante,** Canonical Wick rotations in 3-dimensional gravity, 2009

925 **Sergey Zelik and Alexander Mielke,** Multi-pulse evolution and space-time chaos in dissipative systems, 2009

924 **Pierre-Emmanuel Caprace,** "Abstract" homomorphisms of split Kac-Moody groups, 2009

923 **Michael Jöllenbeck and Volkmar Welker,** Minimal resolutions via algebraic discrete Morse theory, 2009

922 **Ph. Barbe and W. P. McCormick,** Asymptotic expansions for infinite weighted convolutions of heavy tail distributions and applications, 2009

921 **Thomas Lehmkuhl,** Compactification of the Drinfeld modular surfaces, 2009

920 **Georgia Benkart, Thomas Gregory, and Alexander Premet,** The recognition theorem for graded Lie algebras in prime characteristic, 2009

919 **Roelof W. Bruggeman and Roberto J. Miatello,** Sum formula for SL_2 over a totally real number field, 2009

918 **Jonathan Brundan and Alexander Kleshchev,** Representations of shifted Yangians and finite W-algebras, 2008

917 **Salah-Eldin A. Mohammed, Tusheng Zhang, and Huaizhong Zhao,** The stable manifold theorem for semilinear stochastic evolution equations and stochastic partial differential equations, 2008

916 **Yoshikata Kida,** The mapping class group from the viewpoint of measure equivalence theory, 2008

915 **Sergiu Aizicovici, Nikolaos S. Papageorgiou, and Vasile Staicu,** Degree theory for operators of monotone type and nonlinear elliptic equations with inequality constraints, 2008

914 **E. Shargorodsky and J. F. Toland,** Bernoulli free-boundary problems, 2008

913 **Ethan Akin, Joseph Auslander, and Eli Glasner,** The topological dynamics of Ellis actions, 2008

912 **Igor Chueshov and Irena Lasiecka,** Long-time behavior of second order evolution equations with nonlinear damping, 2008

911 **John Locker,** Eigenvalues and completeness for regular and simply irregular two-point differential operators, 2008

910 **Joel Friedman,** A proof of Alon's second eigenvalue conjecture and related problems, 2008

909 **Cameron McA. Gordon and Ying-Qing Wu,** Toroidal Dehn fillings on hyperbolic 3-manifolds, 2008

908 **J.-L. Waldspurger,** L'endoscopie tordue n'est pas si tordue, 2008

907 **Yuanhua Wang and Fei Xu,** Spinor genera in characteristic 2, 2008

906 **Raphaël S. Ponge,** Heisenberg calculus and spectral theory of hypoelliptic operators on Heisenberg manifolds, 2008

905 **Dominic Verity,** Complicial sets characterising the simplicial nerves of strict ω-categories, 2008

904 **William M. Goldman and Eugene Z. Xia,** Rank one Higgs bundles and representations of fundamental groups of Riemann surfaces, 2008

903 **Gail Letzter,** Invariant differential operators for quantum symmetric spaces, 2008